GOVERNMENT

AND HOUSING

GOVERNMENT

AND HOUSING

Developments in
Seven Countries

Edited by

Willem van Vliet--
Jan van Weesep

Volume 36, URBAN AFFAIRS ANNUAL REVIEWS

SAGE PUBLICATIONS
The International Professional Publishers
Newbury Park London New Delhi

For information address:

 SAGE Publications, Inc.
2111 West Hillcrest Drive
Newbury Park, California 91320

SAGE Publications Ltd.
28 Banner Street
London EC1Y 8QE
England

SAGE Publications India Pvt. Ltd.
M-32 Market
Greater Kailash I
New Delhi 110 048 India

Printed in the United States of America

Government and housing : developments in seven countries / edited by
  Willem van Vliet-- and Jan van Weesep.
      p.   cm. — (Urban affairs annual reviews : v. 36)
  Includes bibliographical references.
  ISBN 0-8039-3666-4. — ISBN 0-8039-3667-2 (pbk.)
  1. Housing policy.   I. Van Vliet--, Willem, 1952-   II. Weesep, J.
van.   III. Series.
HT108.U7   vol. 36
[HD7287]
307.76 s—dc20
[363.5'8]   90-32761
                                                                   CIP

FIRST PRINTING, 1990

Sage Production Editor: Kimberley A. Clark

Contents

Acknowledgments

THE PRODUCTION OF THIS VOLUME was prompted by the international research conference "Housing, Policy, and Urban Innovation" held in Amsterdam, June 27-July 1, 1988, under the auspices of the Working Group on Housing and the Built Environment of the International Sociological Association, the Netherlands Organization of Research Institutes in the Field of Housing and Urban Research (GS), and the Dutch national program of urban research (Urban Networks). The conference was sponsored and organized by the Center for Metropolitan Research (CGO) at the University of Amsterdam, the Institute of Geographical Research (IRO) at the University of Utrecht, and the OTB Research Institute for Policy Sciences and Technology at Delft University.

Of the more than 220 papers presented at this conference, a selection was made to provide a combination of perspectives on the changing role of government in housing. An attempt was made to ensure a range of disciplinary approaches to the relevant issues and to include nations that represent diverse political economies. Selected papers were subsequently revised or fully rewritten in response to feedback from the editors. In addition, this volume contains a number of specially invited contributions intended to balance the thematic organization of the book. We are grateful to Blaise Donnelly of Sage Publications for his initiative in developing this project and to Susan Clarke, Dennis Judd, and Gary Tobin, series editors of the Urban Affairs Annual Reviews, for their support.

—Willem van Vliet--
—Jan van Weesep

1

The Privatization and Decentralization of Housing

WILLEM van VLIET--

IN MOST COUNTRIES around the world today, government acknowledges certain responsibilities in assuring citizens agreed-upon minimum living standards, including the provision of decent and affordable housing. The precise nature of these responsibilities, the scope of the standards, and the modes as well as effectiveness of government intervention vary significantly from one country to another. Factors such as political ideology, economic resources, level of technological development, and extent of institution building underlie such variation. A convenient distinction places capitalist market systems, characterized by a minimal role of national government, at one end of the spectrum and socialist redistributive systems, where national government plays a key role, at the other end. These two types of systems feature widely different approaches to housing provision and urban development.

Housing provision in capitalist nations—such as the United States, Canada, Australia, and Japan—has been patterned according to a "free" market model. It treats housing as a commodity, similar to automobiles and refrigerators. Provision of housing is expected to result from the interplay of supply and demand, where transactions are driven by the profit motive of private investors. A fundamental assumption underlying the capitalist model of housing provision is that households that are unable to compete for the more desirable housing stock will occupy units that are vacated and "filter down" as more affluent households move into new housing of higher standards (see, e.g., Tucker, 1989). Government's role is minimal, restricted to actions intended to ensure smooth market functioning.

In reality, however, capitalist systems do not have genuinely "free" markets. The interplay of supply and demand is molded by numerous government interventions. For example, the laying out and paving of streets; the supply of water; sewage disposal; utility provision; material and building standards; zoning ordinances; health, fire, and safety codes; regulation of labor as to wages, hours, and conditions; mortgage insurance; regulation of financial institutions; mortgage foreclosure laws; regulation of tenancy, including eviction, rent level, and even household composition—all of these things are known to be affected by government actions (Marcuse, 1990; see also Otnes, 1988). Government tax expenditures also tend to be highly regressive and to favor owners over renters (see, e.g., Chapter 9 of this volume). The socioeconomic polarization to which government thus contributes reinforces market segmentation, which obstructs the filtering down of housing. Filtering also works imperfectly because of other widespread market distortions, in part reflecting the "risk aversiveness" of (a system driven by) private investors motivated by profit seeking. As access to housing is foremost a function of ability to pay, renters as well as home buyers unable to articulate effective market demands frequently encounter discriminatory practices, and homelessness is a poignant outcome.

By contrast, in centrally planned socialist nations—such as those found in the East Bloc before its current transformation and, for example, China, Vietnam, Cuba, and, until recently, Nicaragua in the Third World—a major theoretical premise of societal organization is that the state distributes costs and benefits, resulting from national functioning and development, equally among all segments of the population. According to this egalitarian ideology, the state must maintain full administrative control over rationally conducted planning, production, management, and consumption processes. In line with these normative principles, housing is viewed, and often legislated, as an entitlement, and construction, distribution, and management of housing are essential state responsibilities.

However, the post-World War II situation presented national authorities in socialist systems with a dilemma: On the one hand, political ideology demanded universal provision of housing by the state against low cost for its citizens (typically 5% or less of household income); on the other hand, economic realities required that financial, material, and human resources would be committed to productive investments with higher returns, chiefly in the heavy manufacturing sector. In this trade-off, the latter generally received priority. Hence the state has frequently had to allow for alternative modalities in order to ameliorate serious housing problems. There has been, for example, spontaneous construction by urban enterprises in need of housing for their employees, and self-help building in rural areas and small towns has contributed to private home ownership rates that are generally much higher than those found in capitalist countries. Privileged positioning of, for example, the party cadre, the intelligentsia, and the military elite in channels of bureaucratic allocation has led to inequalities and, while homelessness is limited, housing standards are relatively low and household crowding is common.

The capitalist and socialist models can fulfill a useful heuristic function in analysis. However, in reality all national housing systems include features of both models and approximate one or the other in different ways. Each has its particular merits and problems. In Turner's (1976) terms, we find that national systems of housing provision commonly involve, in different ways and in different degrees, actions by the public sector (government), the private sector (profit-seeking investors), and the popular sector (users). During the last decade, the interrelationships among these three sectors have changed significantly in many countries worldwide (van Vliet--, 1990b). Although the changes that can be observed are complex and vary among nations, they typically revolve around new roles of national government. Two trends are particularly salient: privatization and decentralization.

In different countries these developments can occur for different reasons, in different forms, and with different ramifications. There is general agreement, for example, that privatization in Britain has been substantially inspired by considerations of political ideology, whereas in the United States, proponents have sought to put more emphasis on the economic benefits of privatization as a pragmatic technical exercise (although certainly not void of ideological motives either—see, e.g., Chapter 7, this volume; see Swann, 1988, for a comparison of Britain and the United States). In Hungary, privatization in housing provision during the 1970s was propelled more strongly by political factors, with economic factors becoming dominant during the 1980s. Expressions of privatization similarly vary. They include the elimination of public programs, the divestiture of public assets to private ownership, deregulation that permits participation by private firms in activities that were previously monopolized by government, and so on. As will be noted below, the implications of privatization and decentralization depend on a variety of factors; there is no simple or uniform outcome. Potential effects include enhanced efficiency and greater cost-effectiveness as well as increased socioeconomic polarization and intensified local conflict.

The chapters in this book were selected to produce further insights into these diverse aspects of the changing government role in housing, subsumed under the broad labels of privatization and decentralization. Thus they illuminate implicated issues of housing policy for welfare states (Britain, Sweden, Israel) and capitalist countries more strongly oriented to private profit making (United States, Australia) as well as socialist nations (Hungary, Yugoslavia). The introductions to the various parts making up the book serve to identify main points emerging from specific chapters and relate them to their particular topical frameworks. The purpose of this chapter is to bring into focus broader questions underlying the changing government role in housing and to situate them in a wider context.

PRIVATIZATION AND DECENTRALIZATION: WHAT, WHERE, WHY, AND HOW?

Recent years have shown a rapid proliferation of the literature on privatization and decentralization. Thousands of publications have appeared in which authors have characterized these developments in a range of specifying terms, and it is easy to get lost in a definitional quagmire. Although it is analytically important to have precise conceptual clarity, it suffices for the present purpose to have a basic understanding of what privatization and decentralization entail in a broad sense. Privatization can be taken to refer to "actions taken by actors legitimately representing the public sector to transfer the hitherto public responsibility for a certain activity or activities away from the public and into the private sector" (Lundqvist, 1988, p. 12). Privatization can also be seen to include "the introduction into the public sector of conditions that typify the private sector" (Swann, 1988, p. 2), for example, profit seeking, cost minimization, and market competition. In both interpretations, privatization is the direction of a process rather than a condition. The denotation is relative rather than absolute.

A different kind of privatization is the withdrawal of individual affective interest and social involvement from the public sphere. In this connection, Young and Willmott (1973) have described how the families they studied began to invest more of their time and money in the private sphere of the family and less in the public sphere of streets and neighborhood as they equipped themselves with larger homes and appurtenant consumer goods. More recently, the decline of involvement in community affairs and the increased pursuit of self-interest inspired a critical assessment by Bellah et al. (1985). Popenoe (1985) has described this situation as "private pleasure, public plight," echoing Galbraith's (1958) contrast of private affluence and public squalor and related observations made by Sennett (1977) about the decline of public culture. Although it is possible to discern related ideological underpinnings (Starr, 1989, p. 18), the concern here is not with the privatization of community life but with the privatization of the functions of government and its representative institutions, with special reference to housing and urban development.

Decentralization, for the present purpose, simply refers to the devolving of national responsibilities to lower levels of government. Here, too, the reality is more complex, as, for example, political decentralization may interact with economic centralization, creating a class of problems that illustrate the importance of contextual considerations (see below).

The well-publicized developments in Britain and their emulation by the United States and other capitalist countries notwithstanding, privatization in housing can be seen in most parts of the world. The chapters on Hungary and Yugoslavia in this book add to a growing literature on the transition of socialist redistributive systems to capitalist market systems (e.g., Lowe & Tosics, 1988; Nee, 1989; Pejovich, 1988; Tosics, 1987; White, 1988). Privatization is also widely evident in Third World countries (e.g., Cook & Kirkpatrick, 1988; Lim & Moore 1989), including countries that are modeled after socialist tenets (e.g., Cheung, 1989; Lee, 1988).

Decentralization is similarly found in capitalist countries (e.g., Dixon, 1989; Kawashima & Stöhr, 1988; Wallis & Oates, 1988), welfare states (e.g., Alterman, 1988; Lindberg & Karlberg, 1988), and the Third World (e.g., King, 1988; Rondinelli, McCullough, & Johnson 1989; U.S. Agency for International Development, 1989). It is also becoming increasingly salient in Eastern Europe, as political and economic reforms continue to unfold and state bureaucracies are being dismantled and replaced by more pluralistic arrangements.

Why the accelerated privatization and decentralization in recent years? Observers point to various reasons. Most prominent among them are economic determinants (recession, fiscal austerity, regional autonomy) and political determinants (ascendancy of conservative ideologies, localism). In the case of the Third World and several East Bloc countries, institutional elements have been identified as major contributing variables, as agencies such as the International Monetary Fund, the World Bank, UNCHS, and U.S. Agency for International Development have made credit and assistance increasingly contingent upon the adoption of market principles.

Although these various economic, political, and institutional factors are documented empirical correlates of privatization and decentralization, they do not offer a ready explanation for these developments. For example, the unremitting attempts at privatization of the social housing stock in Britain during the economic downturn of the 1970s have found but a limited following in other welfare states. From a more theoretical perspective, Lundqvist (1989a), in an examination of the Swedish debate over a 1984 housing tax proposal, concludes that while economic factors may initiate policy actions regarding privatization, ideological, political, and institutional factors are decisive for the content of such action. Using theoretical explanations of welfare-state expansion "in reverse" to explain welfare-state contraction, he attempts to formulate preliminary predictions about when and where certain patterns of privatization will occur (1989b). Likewise, there exist a number of theoretical approaches to decentralization and central-local relationships from statutory, administrative, political, and economic perspectives (e.g., Bollinger, 1983; Gottdiener, 1987; Hansen & Levine, 1988; Hawkins, 1989; Kincaid, 1989; Page & Goldsmith, 1987). Such theoretical work contributes to our understanding of the antecedents of privatization and decentralization. It also helps to clarify how these processes take place.

Privatization can be viewed as a general strategy that proceeds tactically in various domains where its implementation is achieved through the application of a variety of privatization techniques (e.g., Walker, 1988). Lundqvist (1988) distills an emerging consensus from the literature that suggests that activities related to privatization fall into any of three categories: production, financing, and regulation. In this connection, production includes construction, ownership, and management of housing. Financing refers to the funding of housing provision, which may be subsidized or may originate from private sources. Regulation connotes intervention by means of imposing controls, stipulating requirements, and applying rules, all of which serve to restrain freedom of action and structure market processes. Sometimes regulation is ostensibly self-imposed, as when producers respond to perceived consumer preferences (Swann, 1988). Ordinarily, however, regulation is a function of government intervention that reflects the balance of power between contending forces. This same tripartite classification of production, financing, and regulation can be useful as well in examining the decentralization of activities related to housing provision.

As is obvious from the chapters that follow, any given component of a housing system is embedded in a complex matrix in which privatization and decentralization can form multiple combinations. For example, in various countries (although by no means all) public building has been severely curtailed or has ceased altogether, and government policies have promoted the sale of public housing units to tenants or other private parties (e.g., in Britain, the United States, and Israel). These actions represent privatization in the production sphere. In the finance sphere, they are often concurrent to a decline in subsidies for social housing, particularly object subsidies aimed at the supply side, with a concomitant increase of the burden on the private and nonprofit sector to raise requisite funding (see, e.g., Chapters 6, 9, 10, 11, and 12). However, privatization and decentralization in both production and financing have frequently been accompanied by *in*creased regulation by national government intended to ensure that actions at the local level and in the private sector comply with stated national policy goals. A case in point is the intervention by the conservative Thatcher administration to enforce "right to buy" legislation over the will of local authorities controlled by the Labour party (Forrest & Murie, 1987, pp. 25-29; see Laws, 1988, for a similar example from Canada).

It is beyond the scope of this short chapter to review the many specific and diverse ways in which privatization and decentralization manifest themselves in housing and urban development. Such a review can be valuable (see U.S. General Accounting Office, 1989, for an example of a useful inventory), but it becomes relevant only in the context of outcomes attributable to privatization and decentralization. In other words, notwithstanding all the attention afforded privatization and decentralization, their ramifications are what must concern us most.

PRIVATIZATION AND DECENTRALIZATION: SO WHAT?

Privatization and decentralization both find ardent advocates and fervent opponents. There is a dearth of systematically documented effects to support either position without qualification, not in the least because of methodological and conceptual difficulties (e.g., problems in pre- and postmeasurement, unit of analysis, sample size, selection of independent and dependent variables and their operationalization, spurious correlations). However, some tentative conclusions can be drawn.

Regarding privatization, the line of reasoning in the economic realm is typically as follows: Deregulation removes constraints on market transactions, which, if accompanied by private ownership of capital and means of production, will result in increased competition, which would then translate into cost savings and greater efficiency. Two large international surveys, discussed by Swann (1988), provide mixed evidence. Millward (1982) found no broad support for private enterprise superiority and Borcherding, Pommerehne, and Schneider (1982), citing more than 50 studies from five countries, conclude that conditions of competition rather than private ownership improve performance. In a recent comparative analysis of housing provision, Nesslein (1988) concludes that intervention by national government, characteristic of welfare states as found in Scandinavia and Western Europe, has not led to higher *overall* national investment in housing, nor lowered *overall* national costs, in comparison with countries whose housing provision is patterned after the market model. Economic desiderata, like those considered in these studies, form an integral ingredient of any housing policy. However, a balanced assessment of privatization requires a broader-based approach.

If housing is treated as an economic commodity, as is typically the case in capitalist market systems, increasing production efficiency and reducing costs clearly emerge as major criteria for policy formulation. However, in welfare states and socialist systems, where housing is viewed more as an entitlement, these criteria do not provide the chief guidelines for policy, but tend to be considered within the more important context of goals to reduce inequities in access to housing and to lower *individual* costs such that decent housing is affordable also to segments of the population unable to compete effectively in the market. In this connection, questions arise regarding the differential impacts of privatization on economic versus more socially oriented goals in housing provision. Does privatization lead to reduced government expenditure? What are its redistributive effects?

In Chapter 7 of this volume, Silver indicates how privatization of the U.S. public housing stock, as in Britain, has failed to produce financial benefits. The British case is of special interest, as it shows how government housing expenditure was shifted rather than cut. On the face of it, the national housing budget appears to have declined. However, alternative computations by Forrest and Murie (1987) reveal how the seeming reduction in housing expenditure of £1,318 million between 1979-80 and 1984-85 was, in fact, more than offset by the increase of £2,020 in housing benefits (disguised as social security expenditures). Forrest and Murie, therefore, argue that, notwithstanding the government's privatization policies and apparent budget cutbacks, actual public housing expenditure was not reduced but rather was reoriented toward consumers of housing within a framework that strongly favors private home owners over renters. This observation brings up the more general point of the redistributive effects of privatization.

During the past decade, there has been considerable socioeconomic polarization in housing in, for example, the United States, Canada, and Britain, particularly across tenure lines (Van Vliet--, 1990a). The result has been a residualization of the rental sector, typically already marginal to economic and political processes. Privatization of housing provision is likely to intensify this polarization. Greater reliance on market processes makes the economically disadvantaged more vulnerable, since they are not in a position to compete effectively for quality housing. Similarly, while the effects of the deregulation of the mortgage industry on housing affordability remain unclear, buying a home may become more difficult for some population groups: Deregulation creates greater potential for discriminatory practices by lending institutions against the poor and minorities, based on perceived or alleged investment risks. Deregulation can also affect a buyer's ability to keep a home, as it interacts with economic hardships. For example, Heisler and Hoffman (1987) found that delinquent home owners were less successful in negotiating forbearance with their lenders in the type of nonlocal, competitive mortgage markets that have developed under deregulation than were home owners with locally held mortgages.[1]

Relevant as well in an examination of the redistributive consequences of privatization is the fast-rising significance of multinational corporations (van Vliet-- & Burgers, 1987). The internationalization of production and the greatly increased transnational mobility of capital have led to disinvestment and severe economic slump in some locations and rapid growth with tight housing markets and inadequate community infrastructure in others, thus producing major regional disparities. Further, in tandem with these processes, there is a centralization of the global management and production service functions of multinational corporations in major cities. The accompanying work force requisites induce polarization across skill and income levels (Sassen, 1988). This directly impinges on access to housing as accumulated capital is mobilized for superior market position and speculative investment in built environment, as is frequently the case with gentrification. Consequences of reduced affordability suggest a stronger, not a weaker, government role.[2]

There are many more examples of similar consequences of the retreat by government from its traditional responsibilities in housing provision. Significant among them are expiring government subsidies of social housing in the Federal Republic of Germany and the United States, which, if no corrective steps are taken, will result in drastic losses of units for low-income households (see, e.g., Chapters 6 and 9, this volume; see also Jaedicke & Wollmann, 1990). It is not possible to detail these and other examples here; it suffices to summarize that observed developments indicate little or no economic benefit deriving from the privatization of housing, whereas its redistributive effects are likely to generate inequities and exacerbate polarization.[3]

Decentralization is frequently justified on the grounds that local government can identify community needs, put them in priority order, and address them in appropriate ways better than central government can. This premise often becomes problematic in practice, as local government lacks the resources necessary to address the problems with which it is confronted.[4] A case in point is the problem of homelessness, which stems in large part from economic restructuring and housing policy shifts that transcend local jurisdictions. In many cities, government has responded by using its limited funds as leverage in complicated patchworks of public-private partnerships (e.g., Brindley & Stoker, 1987; Catanese, 1984; Law, 1988). There is also a growing practice of linking permits for urban development to the provision of housing for moderate- and low-income households (e.g., Goetz, 1989; see LeGates, 1988, for a discussion of local backlash actions undertaken in response to housing privatization in the United States). Decentralization thus helps set a context for privatization at the local level.

The U.S. Department of Housing and Urban Development actively seeks to promote these types of approaches that rely on local and private sector initiatives. In its *Directory of Official U.S. Projects in the International Year of Shelter for the Homeless*, the agency made clear that its criterion in selecting award-winning projects was intended "to identify exemplary *local* activities that explore new ways and means of improving the shelter and neighborhoods of low-income families through use of *local* initiatives that involve the *private* sector" (emphasis added). However, privatization and decentralization impose serious constraints on the scope and orientation of professional activities (see, e.g., Dear, 1989; Yanay, 1989), and it is unrealistic to expect local, private sector solutions to housing problems whose local manifestation is anchored in supralocal private sector interests (alluded to also in the earlier discussion of the ascendance of multinational corporations).

Unless undertaken with proper safeguards, instituted by national government, decentralization indubitably diminishes building activity for low-income households and undermines maintenance, contributing to deteriorating housing conditions and shortages. There is a likely negative impact on distribution as well, since decentralization typically gives local officials more autonomy over fewer resources, making their actions more susceptible to the vicissitudes of market processes and thus creating greater potential for local conflict (Clarke, 1989). Further, as is brought out by Nord in Chapter 4 of this volume, decentralization also increases the potential for inequities among different localities. A recent comparative analysis of interlocal cooperation in a nationally representative sample of 26 U.S. metropolitan areas found that, in the absence of sanctions or incentives from higher authorities, communities were able to cooperate when the benefits were shared or the costs hidden, but were unable to do so otherwise (Wrightson, 1986), suggesting greater effectiveness of intervention by central authorities.

Brief mention has already been made of the interrelationships among aspects of privatization and decentralization. These connections are important in various ways. For example, the ways and extent that housing provision can be privatized are conditioned in part by the structure of government and the institutional organization of housing provision. Federally based systems, as found in the United States and Australia, have elected political decision-making entities at the state and local levels with considerable jurisdictional autonomy and responsibility in housing and urban development. While national government may sanction privatization in general, specific initiatives are developed at lower levels of government. In welfare states, such as Sweden, and in centrally planned socialist countries, like Hungary, national government plays a more prominent role, as it can mandate privatization that is binding on lower levels. An implication of the linkages between privatization and decentralization is that communities in decentralized systems are less able to protect themselves against adverse effects of privatization on, for example, affordability, as they are less insulated from such effects by national intervention and their redistributive capacities are inherently limited. Furthermore, under administrative devolution, accountability vis-à-vis the attainment of housing goals may become problematic (see Carroll, 1989).

As noted, the extent of (de)centralization forms a contingent condition for privatization. Opportunities for and implications of privatization also vary according to other contexts. In the United States, public housing accounts for only 1.5% of the national housing stock and lacks a powerful constituency. There is little to privatize and relatively little that stands in the way. By contrast, in the Netherlands, 43% of the housing stock is in the social sector, which is strongly represented by large, influential organizations. There is much to privatize, but many vested interests act as barriers (see Priemus, 1990).

In Eastern Europe, proponents of privatization also face a problematic institutional context. Mechanisms necessary for effective functioning of housing markets, such as mortgage banks and real estate agencies, are absent or insufficiently developed. Further, funds for loans are inadequate and private capital is extremely limited. The "second" or "informal" economy remains essential (Sampson, 1988; see also Portes & Böröcz, 1988). Under these circumstances, self-help practices, analogous to those found in the Third World, can be expected to be of continued significance. Although on a different basis, in Western countries privatization also appears to engender a greater role for the family in housing provision (Forrest, 1988; Tosi, 1989) and for the nonprofit sector in urban development more generally (see, e.g., Chapter 12, this volume; Laws, 1988; Randall & Wilson, 1989).

CONCLUSION

It is not possible in this brief chapter to cover the numerous issues related to the privatization and decentralization of housing and to do justice to their complexity. However, many of them are ably treated in the more general literature on these topics, and the chapters that follow are rich in specific insights regarding their role in housing. The purpose here has not been to provide comprehensive or detailed coverage, but to identify several underlying questions of broader significance. These questions relate particularly to the consequences of privatization and decentralization. It is their effects on the availability, affordability, and quality of housing that are first and foremost of interest. It would be a mistake, however, to focus on these effects at the exclusion of other concerns. It is also important to emphasize that privatization and decentralization occur in conjunction with and in the context of other processes, briefly referred to above, that have an impact on housing as well. Moreover, current arrangements of activities across the spectrum of housing provision and urban development often involve an intertwining of responsibilities for various combinations of parties, for example, in financing, development of land, construction, distribution, and management of housing. These blurred boundaries negate simplistic dualist conceptions of public-private and central-local relationships. Awareness and careful assessment of this reality should inform innovative approaches to the extant problems.

NOTES

1. In the United States, the federal government is undertaking an unprecedented bailout of the fledgling savings-and-loan industry. The crisis resulted from a combination of poor judgment and outright fraud, made possible by deregulation, and is estimated by the General Accounting Office to cost taxpayers about $500 billion, or $2,000 per capita. Legislation adopted to control the damage of the fiasco and prevent a recurrence mandates mergers and imposes tightened operating rules.

2. An example in this connection is the intervention in September 1987 by the federal government in Australia, which sought to curtail the scope of operations of foreign capital in the housing market out of a concern for inflationary effects on house prices (Rimmer, 1989). It should be noted that a stronger role for government does not necessarily mean more regulation. A recent study is suggestive of the limits of regulatory measures in the provision of affordable housing (Dalton, 1989). Policy implementation can involve an array of instruments, and their effectiveness depends, in part, on the judiciousness of their application.

3. Although acknowledging this possibility, Stren (1989) points out how in Africa (and, evidence indicates, in much of the Third World) state actions have frequently reinforced existing political and class divisions, while at the same time urban infrastructure and services have deteriorated to a point where the productive function of cities is in jeopardy. In this light, privatization may have positive effects as it helps to ameliorate serious urban problems; the poor have a better chance of improving their situation through their involvement in this process than through intervention by the state, which in the past has been largely inimical to their needs.

4. An interesting trend can be noted in the Netherlands, which has actively pursued a policy of decentralization, using relatively generous formulae for revenue sharing. However, from 1982 through 1988 the number of municipal employees in the Netherlands decreased by 4.7%, whereas the number of national government employees increased by .5% (Wereld, 1988).

REFERENCES

Alterman, R. (1988, Fall). Implementing decentralization for neighborhood regeneration: Factors promoting or inhibiting success. *Journal of the American Planning Association*, pp. 454-469.

Bellah, R. N., et al. (1985). *Habits of the heart: Individualism and commitment in American life*. Berkeley: University of California Press.

Boelhouwer, P. J., & van Weesep, J. (1988). On shaky grounds: The case for the privatization of the public housing sector in the Netherlands. *Netherlands Journal of Housing and Environmental Research, 3*(4), 319-333.

Bollinger, S. J. (1983). The historic and proper place of central governments in urban redevelopment: The U.S. view. *Urban Law and Policy, 6*, 53-63.

Borcherding, T. E., Pommerehne, W. W., & Schneider, F. (1982). Comparing the efficiency of private and public production: The evidence from five countries. *Zeitschrift für National Ökonomie* [Supplement 2], pp. 127-156.

Brindley, T., & Stoker, G. (1987). The privatization of housing renewal: Dilemmas and contradictions in British urban policy. In W. van Vliet-- (Ed.), *Housing markets and policies under fiscal austerity* (pp. 33-47). Westport, CT: Greenwood.

Carroll, B. (1989). Administrative devolution and accountability: The case of the non-profit housing program. *Canadian Public Administration, 32*(3), 345-366.

Catanese, A. J. (1984). *The politics of planning and development.* Beverly Hills, CA: Sage.

Cheung, S. N. S. (1989). Privatization vs. special interests: The experience of China's economic reforms. *Cato Journal, 8*(3), 585-596.

Clarke, S. E. (1989). The political implications of fiscal policy changes. In S. E. Clarke (Ed.), *Urban innovation and autonomy* (pp. 236-251). Newbury Park, CA: Sage.

Cook, P., & Kirkpatrick, C. (Eds.). (1988). *Privatisation in less developed countries.* New York: St. Martin's.

Dalton, L. C. (1989, Spring). The limits of regulation: Evidence from local plan implementation in California. *Journal of the American Planning Association*, pp. 151-168.

Dear, M. (1989). Privatization and the rhetoric of planning practice. *Environment and Planning D: Society and Space, 7*, 449-462.

Dixon, J. E. (1989). From centralism to devolution: Reforming local government in New Zealand. *Regional Studies, 23*(3), 267-270.

Forrest, R. (1988, September 16-19). *Between state and market: Privatisation, family resources and aspects of early household formation in European housing systems.* Paper prepared for the international workshop, Housing Between State and Market, Inter-University Centre, Dubrovnik.

Forrest, R., & Murie, A. (1987). Fiscal reorientation, centralization, and the privatisation of council housing. In W. van Vliet-- (Ed.), *Housing markets and policies under fiscal austerity* (pp. 15-31). Westport, CT: Greenwood.

Galbraith, J. K. (1958). *The affluent society.* Boston: Houghton Mifflin.

Goetz, E. (1989). Office-housing linkage in San Francisco. *Journal of the American Planning Association, 55*(1), 66-77.

Gottdiener, M. (1987). *The decline of urban politics: Political theory and the crisis of the local state.* Newbury Park, CA: Sage.

Hansen, M. G., & Levine, C. H. (1988). The centralization-decentralization tug-of-war in the new executive branch. In C. Campbell & B. G. Peters (Eds.), *Organizing governance: Governing organizations* (pp. 255-282). Pittsburgh: University of Pittsburgh Press.

Hawkins, R. B., Jr. (1989). Linking constitutional reform to local self-governance. *Journal of State Government, 62*(1), 31-33.

Heisler, B. S., & Hoffman, L. M. (1987). Keeping a home: Changing mortgage markets and regional economic distress. *Sociological Focus, 20*(2), 227-241.

Jaedicke, W., & Wollmann, H. (1990). The Federal Republic of Germany. In W. van Vliet-- (Ed.), *The international handbook of housing policies and practices.* Westport, CT: Greenwood/Praeger.

Kawashima, T., & Stöhr, W. (1988). Decentralized technology policy: The case of Japan. *Environment and Planning C: Government and Policy, 6*, 427-439.

Kincaid, J. (1989). A proposal to strengthen federalism. *Journal of State Government, 62*, 36-45.

King, D. Y. (1988). Civil service policies in Indonesia: An obstacle to decentralization? *Public Administration and Development, 8*, 249-260.

Kitay, M. G. (1987). Public-private co-operation for the development of low-income shelter. *Habitat International, 11*(1), 29-36.

Law, C. (1988). Public-private partnership in urban revitalization in Britain. *Regional Studies, 22*, 446-452.

Laws, G. (1988). Privatisation and the local welfare state: The case of Toronto's social services. *Transactions of the Institute of British Geographers, 13*, 433-448.

Lee, Y. F. (1988). The urban housing problem in China. *China Quarterly, 115*, 387-407.

LeGates, R. T. (1988). The local government backlash against federal housing privatization in the United States. *Built Environment, 14*(3/4), 190-200.

Lim, G. C., & Moore, R. J. (1989). Privatization in developing countries: Ideal and reality. *International Journal of Public Administration, 12*(1), 137-161.

Lindberg, G., & Karlberg, B. (1988). Decentralisation in the public housing sector in Sweden. *Scandinavian Journal of Housing and Planning Research, 5*, 1-15.

Lowe, S., & Tosics, I. (1988). The social use of market processes in British and Hungarian housing policies. *Housing Studies, 3*(3), 159-171.

Lundqvist, L. J. (1988). Privatization: Towards a concept for comparative policy analysis. *Journal of Public Policy, 8*, 1-19.

Lundqvist, L. J. (1989a). Economics, politics, and housing finance: What determines the choice between privatization and collective redistribution? *Scandinavian Housing and Planning Research, 6*, 201-213.

Lundqvist, L. J. (1989b). Explaining privatization: Notes towards a predictive theory. *Scandinavian Political Studies, 12*(2), 129-145.

Marcuse, P. (1990). The United States of America. In W. van Vliet-- (Ed.), *The international handbook of housing policies and practices*. Westport, CT: Greenwood/Praeger.

Millward, R. (1982). The comparative performance of public and private ownership. In L. Roll (Ed.), *The mixed economy* (pp. 58-93). London: Macmillan.

Nee, V. (1989). A theory of market transition: From redistribution to markets in state socialism. *American Sociological Review, 54*(5), 663-681.

Nesslein, T. S. (1988). Housing in the welfare state: Have government interventions raised housing investment and lowered housing costs? *Urban Affairs Quarterly, 24*(2), 295-314.

Otnes, P. (1988). Housing consumption: Collective systems service. In P. Otnes (Ed.), *The sociology of consumption* (pp. 119-138). Oslo: Solum Forlag A/S.

Page, E. C., & Goldsmith, M. J. (Eds.). (1987). *Central and local government relations: A comparative analysis of West European unitary states*. London: Sage.

Pejovich, S. (1988). Freedom, property rights and innovation in socialism. In M. A. Walker (Ed.), *Freedom, democracy and economic welfare* (pp. 323-354). Vancouver, BC: Fraser Institute.

Popenoe, D. (1985). *Private pleasure, public plight: American metropolitan community life in comparative perspective*. New Brunswick, NJ: Transaction.

Portes, A., & Böröcz, J. (1988). The informal sector under capitalism and state socialism: A preliminary comparison. *Social Justice, 15*(3/4), 17-28.

Priemus, H. (1990). The Netherlands. In W. van Vliet-- (Ed.), *The international handbook of housing policies and practices*. Westport, CT: Greenwood/Praeger.

Randall, R., & Wilson, C. (1989). The impact of federally imposed stress upon local-government and nonprofit organizations. *Administration & Society, 21*(1), 3-19.

Rimmer, P. J. (1989). Japanese construction contractors and the Australian states: Another round of interstate rivalry. *International Journal of Urban and Regional Research, 13* 404-424.

Rondinelli, D. A., McCullough, J. S., & Johnson, R. W. (1989). Analysing decentralization policies in developing countries: A political-economy framework. *Development and Change, 20*(1), 57-88.

Sampson, S. (1988). "May you live only by your salary!": The unplanned economy in Eastern Europe. *Crime and Social Justice, 8.*

Sassen, S. (1988). *The mobility of labor and capital.* New York: Cambridge University Press.

Sennett, R. (1977). *The fall of public man.* New York: Knopf.

Starr, P. (1989). The meaning of privatization. In S. B. Kamerman & A. J. Kahn (Eds.), *Privatization and the welfare state.* Princeton, NJ: Princeton University Press.

Stren, R. E. (1988). Urban service in Africa: Public management or privatisation? In P. Cook & C. Kirkpatrick (Eds.), *Privatisation in less developed countries* (pp. 217-247). New York: St. Martin's.

Swann, D. (1988). *The retreat of the state: Deregulation and privatization in the UK and US.* Ann Arbor: University of Michigan Press.

Tosi, A. (1989, June 27-July 1). *Housing models and informal practices in industrialized countries: The role of family networks.* Revision of a paper presented at the International Research Conference on Housing, Policy, and Urban Innovation, Amsterdam.

Tosics, I. (1987). Privatization in housing policy: The case of western countries and that of Hungary. *International Journal of Urban and Regional Research, 11*(1), 61-78.

Tucker, W. (1989, Fall). Home economics: The housing crisis that overregulation built. *Policy Review,* pp. 20-24.

Turner, J. F. C. (1976). *Housing by people: Towards autonomy in building environments.* New York: Pantheon.

U.S. Agency for International Development, Office of Housing and Urban Development Programs. (1989). *Annual report fiscal year 1989.* Washington, DC: Author.

U.S. General Accounting Office. (1989). *Partnership projects: Federal support for public-private housing and development efforts* (GAO/PEMD-89-25FS). Washington, DC: Government Printing Office.

van Vliet--, W. (1990a). Cross-national housing research: Analytical and substantive issues. In W. van Vliet-- (Ed.), *The international handbook of housing policies and practices* (pp. 1-82). Westport, CT: Greenwood/Praeger.

van Vliet--, W. (Ed.). (1990b). *The international handbook of housing policies and practices.* Westport, CT: Greenwood/Praeger.

van Vliet--, W., & Burgers, J. (1987). Communities in transition: From the industrial to the postindustrial era. In I. Altman & A. Wandersman (Eds.), *Neighborhood and community environments* (pp. 257-290). New York: Plenum.

Walker, M. A. (Ed.). (1988). *Privatization: Tactics and techniques.* Vancouver, BC: Fraser Institute.

Wallis, J. J., & Oates, W. E. (1988). Decentralization in the public sector: An empirical study of state and local government. In H. S. Rosen (Ed.), *Fiscal federalism: Quantitative studies* (pp. 5-32). Chicago: University of Chicago Press.

Wereld Omroep. (1988, August 22). [News commentary]. Netherlands World Radio.

White, G. (1988). State and market in China's socialist industrialisation. In G. White (Ed.), *Developmental states in East Asia* (pp. 153-192). London: Macmillan.

Wrightson, M. (1986). Interlocal cooperation and urban problems: Lessons for the new federalism. *Urban Affairs Quarterly, 22*(2), 261-275.

Yanay, U. (1989). Limits to professional practice in decentralized systems. *Social Policy and Administration, 23*(1), 48-59.

Young, M., & Willmott, P. (1973). *The symmetrical family.* New York: Pantheon.

PART I

The Decentralization of Housing

Introduction

CAROLYN TEICH ADAMS

WITHIN THE WORLD'S advanced capitalist economies, one finds few
advocates these days for increasing national government intervention
in the provision of housing. At the same time that they were adopting
private market strategies in the late 1970s and 1980s, governments in
Europe and North America were shifting responsibility for housing
programs to the local level. Admittedly, not all nations were rushing to
decentralize at the furious speed of the Reagan administration in the
United States. Yet by the end of the decade it seemed clear that the
postwar world had seen a marked slowing in the expansion of the
national welfare state in the West, with the result that local communities
would shoulder greater burdens for housing and social services.

Doubtless some proponents of decentralization are inspired by no
more complex a motive than the desire to cut national expenditures.
Others, however, are driven by a sense that the growth of the welfare
state has inflated the power of national government and sapped the
vigor of local authorities. Critics of the welfare state have argued that
too much centralization of political authority has accompanied the
massive increases in central government budgets during the past 40
years, transforming local authorities into mere agents of national purpose, weakening historic traditions of local self-government, and substituting national choices for local ones.

Not only has central power undermined local democracy, the de-
centralizers argue, it has also led to policy failures when central decision makers impose standardized programs without regard to local
conditions and without sensitivity to the differential impacts of their
programs in different local settings. Nowhere is this danger more
apparent than in the housing field; local market conditions often determine whether particular policy approaches will succeed or fail. The

interplay between local market conditions and housing policies is emphasized by two of the chapters that follow in Part I. Writing about the United States, Feldman and Florida note the differential impact of federal postwar housing policies on inner-city housing markets versus suburban markets. Without intending to do so, the creators of the federal mortgage insurance programs hastened the decline of inner-city residential areas while fostering growth in newer suburban communities. Paris's account of Australian housing intervention gives similar attention to the importance of local conditions, going so far as to create a typology of the different local housing market "types" represented by different localities: inner city, outer suburb, country town, resort community, and so on. The typology is intended to alert policymakers to the different dynamics at work in the housing markets within the various community types, each of which has a distinctive set of social and political relations that influence housing production, financing and tenure. A point made by both chapters is that to address a wide variety of submarkets with standardized policies is to invite failure in a large number of them.

Yet, while local variations lead some analysts to propose greater decentralization, they lead others to defend centralized responsibility for housing programs. Surveying the disparities in the economic fortunes of various communities, defenders of a strong national role reason that only the national government is capable of redressing the imbalances that market forces generate. They see national withdrawal as especially problematic given the increasing mobility of international capital, whose cycle of investment and discipline investment has spatial consequences that are increasingly uneven.

New forms of industry are seeking locations outside of established cities and towns. While capital is shifting to new and growing regions, older industrial regions are seeing an erosion of their employment bases, accompanied by losses in housing investments. Many of the declining regions contain concentrations of low-income households. In countries such as West Germany, the Netherlands, Britain, and the United States, immigrants, minorities, and guest workers are clustered in aging cities where the older housing stock is located, as are the single elderly and other low-income groups. These areas' housing needs are likely to be greater than those of more affluent communities, while their local tax capacities are typically more limited.

The three chapters in Part I all address problems inherent in local government's role in housing production and distribution in advanced capitalist systems. In all three cases, governmental housing functions are shared by central and local authorities. However, local government institutions in the three nations possess widely varying degrees of autonomy vis-à-vis higher levels of government. The greatest degree of autonomy is seen in the United States, where local governments rely most heavily on their own resources and exercise the widest scope of discretion over programs. Somewhat less independence is exercised by Swedish local governments, which rely more heavily on national subsidies and act in some respects as agents of the national government. Australian local governments are the weakest of all, historically performing only a narrow range of functions while the stronger state-level governments assume primary responsibility for most public services in cities and towns.

Taken as a group, the chapters in Part I raise two crucial questions that policymakers must address if they are to design the optimal role for local government in housing provision:

- *Question 1:* How can local governments effectively cope with wider economic and political forces affecting their local housing markets?

This question lies at the heart of the first two chapters—one treating the United States and the other analyzing some housing initiatives in Australia. Feldman and Florida illuminate the interconnections between the housing sector and the wider economy, showing how housing development since World War II has both influenced and been influenced by large-scale economic trends. Federal interventions, they show us, influenced local housing markets not only directly through housing policies, but also indirectly through policies to manage the economy by stimulating consumer demand. They note that in the postwar period the United States absorbed much of its growing productive capacity by building suburban residential communities, thereby establishing a style of household consumption that generated demand for more automobiles, more appliances, and generally more household goods. While Feldman and Florida stop short of claiming that federal policies to encourage suburbanization were consciously motivated by a desire to bolster consumer demand, they do show how well such housing policies fit postwar planners' general goal of managing aggregate demand. Their account of housing's contribution to national economic management shows that local housing officials who seek to control development within their boundaries confront not only market forces, but the combined impact of markets and national policies. Paris makes a similar point about the pervasive influence of Australian national government on local housing conditions, noting that policies governing taxes, mortgage interest rates, the regulation of financial institutions, and even immigration laws all have "systemic effects on housing." Moreover, the accelerating economic competition among state-level governments in Australia means that state officials in the 1980s have been willing to override local development preferences in order to attract new investment. Confronting the dynamics of national and international capital markets, as well as the sometimes incompatible goals of higher levels of government, local officials operate within serious constraints. Strategies to decentralize housing policy must acknowledge these constraints and design policies accordingly.

- *Question 2:* Can local governments successfully engage in redistributive housing programs?

The literature on intergovernmental relations in welfare states suggests that social welfare programs that redistribute resources, taxing the "haves" in order to subsidize the "have-nots," should be assigned to higher-level governments rather than to local authorities. This gives such programs the widest possible tax base and reduces the incentives that business firms and individual citizens might have to escape the burden of taxes by simply moving to another jurisdiction. In contrast, local governments are well suited to provide basic "housekeeping" services (e.g., police, fire, sanitation, water, and sewer systems) and to engage in activities that strengthen the local economy, improve the environment for residents and businesses, create jobs, and generate additional tax revenues. The trouble with applying this simple rule of thumb to housing is that housing programs do not fall neatly into one category; they serve *both* social welfare and development functions. On one hand, the distribution, standards, and costs of housing are welfare concerns. On the other hand, local authorities typically desire to control land development within their borders, and frequently their motive for intervening in housing markets is to stimulate the economy and promote employment.

The Swedish case, which is the subject of the third chapter in this section, illustrates how one national system successfully balances central and local responsibility for the redistributive and developmental aspects of housing policy. Under Swedish policy adopted after World War II, local governments play a central role in planning and constructing the majority of all new rental housing through their operation of municipal nonprofit housing corporations. Local planners exercise a virtual monopoly over decisions affecting residential development; not only do they decide how all land within the municipality is to be used, but they also have rights of expropriation and first refusal on land, the right to assemble vast undeveloped tracts in land banks, and access to government loans to cover virtually all construction costs. Thus the local government is firmly in control of the decisions that affect housing's developmental impacts.

At the same time, the redistributive dimension of Swedish housing policy is reflected in its housing allowances, or cash grants, to housing consumers to ensure that they need not pay excessive proportions of their monthly income for housing. Nord's chapter focuses on one of the two major forms of cash subsidy—local housing allowances for the elderly, or LHAE (the other major form being the housing allowance for families with children and low incomes). As a form of welfare support, the responsibility for the LHAE is not borne by local authorities alone, but shared by local and national government. The central government makes rules governing the criteria for allocation, the local government sets the benefit standard, and both levels contribute funds. Moreover, Sweden's municipalities also receive tax equalization grants from the national government that are paid to localities with small tax bases in order to bring their resources up to levels comparable to more affluent jurisdictions. Hence local officials in Sweden do not face the prospect of having to limit benefits to only those that local tax revenues can support. They are free to set benefit levels more in response to need than in response to cost.

Nord's chapter shows, however, that even where local officials are able to share the cost of welfare benefits with national agencies, they do not necessarily set benefit levels in response to socioeconomic need, as defined by local housing costs or population characteristics. Interestingly, the nature of local administrative structure seems to play a large role in determining benefit levels. The single most important predictor of various municipalities' benefit levels for the housing allowances for the elderly, Nord finds, is the type of administrator in charge of the program. Social service professionals appear to set higher benefit levels than do professionals in finance, real estate, or central administrative departments. This finding suggests that devolving responsibility to local government does not invariably lead to more liberal policies.

Paris's account of Australian housing initiatives makes a similar point. He observes that local governments in Australia are seen by national policymakers as too conservative and unlikely to promote significant social change (a situation that parallels U.S. intergovernmental politics in the 1960s, when the War on Poverty funded community groups instead of local governments because municipal officials were too interested in protecting the status quo). National housing bureaucrats in Australia appear to have closer ties to, and more sympathy for, the community sector than do local officials.

Taken together, these chapters suggest that there may be a trade-off in housing policies between local officials' autonomy vis-à-vis national government and their autonomy vis-à-vis market forces. Where local housing programs are more dependent on the national treasury for support, as in Sweden, local officials are not so constrained by market forces. They need not fear that embarking on large-scale programs, particularly redistributive ones, will increase local tax burdens so much that new businesses and residents will be discouraged from moving in. In countries where lower-level governments are less dependent on national subsidies, as in Australia and the United States, local officials avoid costly programs that would create disproportionate tax burdens within their jurisdictions, when compared with other jurisdictions. In that sense, their relative independence vis-à-vis national government makes them more captive to market forces.

2

Economic Restructuring and the Changing Role of the State in U.S. Housing

MARSHALL M. A. FELDMAN
RICHARD L. FLORIDA

THE GREAT DEPRESSION triggered a series of institutional reforms that became the cornerstone of U.S. housing policy after World War II. Housing was crucial to the postwar expansion and had important social and political ramifications. For the first time, more households lived in suburbia than in central cities or rural areas, and owner occupancy was the most common housing tenure. But housing consumption was uneven, and the housing system systematically excluded some groups. Housing differences both reflected and cut across social cleavages rooted in class, race, and labor market position. Housing added another, largely spatial, prism through which social divisions were refracted (Florida & Feldman, 1988).

Starting in the mid-1960s, economic stagnation set in and major economic restructuring followed. This prompted a series of changes in the housing system. As housing began to play a less central economic role, it also became less central politically. These forces combined to set off another wave of institutional change, resulting in privatization of housing and less active federal participation in housing.

This chapter examines these changes. Its central theme is the decline of housing-induced mass consumption as a key component of the postwar political economy. The chapter begins with an overview of housing in the postwar political economy, followed by a discussion of housing's role in the crisis that began in the mid-1960s. The next section discusses the resulting transformation of housing. Major points are summarized with a general interpretation in the final section.

HOUSING IN THE
POSTWAR POLITICAL ECONOMY

Postwar growth in the United States was driven by military spending and suburbanization (Davis, 1984; Florida & Feldman, 1988). In contrast to the social democratic paths charted by most Western European countries, the United States had a highly segmented system of labor relations and very little public redistribution. Instead of public social spending, privatized consumption made possible by high wages provided the main source of effective demand. Income levels were determined on an industry-by-industry basis through collective bargaining. In many industries unions accepted bargaining agreements tying wages to productivity gains, and this allowed workers' real incomes to rise without directly impinging on profits. Other groups, particularly minorities and women, found themselves outside the main current, too weak to alter their own situations and reap more of the benefits of postwar growth. The resulting variations in income and conditions of employment found expression in variegated consumption patterns that further sharpened social divisions.

Housing was central to this process. Through a complex variety of push and pull factors, large numbers of persons embraced suburban living. Suburbanization, and associated consumption patterns, replaced other forms of living with a commodity-intensive form without necessarily raising living standards. Mass suburbanization induced demand, thereby absorbing productivity gains generated by postwar production arrangements. This precluded any sharp reduction in working hours, even as productivity rose.

EFFECTIVE DEMAND

Federal policies implemented during the New Deal allowed housing to become a key element of privatized consumption. On one hand, these policies set the stage for massive suburbanization. Financial reforms made home ownership feasible for a large proportion of the population by defraying the costs of home purchase. As implemented, these reforms favored the detached, single-family, suburban house. Along with federal transportation policies, they accelerated metropolitan decentralization and stimulated demand for housing, automobiles, and related commodities. More than in any other country, these activities in the United States formed the basis for postwar growth. On the other hand, rather than encourage equality in housing, federal policies favored middle-income households and established distinct housing markets.

Postwar housing therefore reflected labor market segmentation while adding spatial and social segmentation in housing as another dimension of social differentiation.

LABOR COSTS AND RESOURCES

Aglietta (1979) and others argue that low-cost postwar housing kept labor costs down. This is questionable. Federal policies made owner-occupied suburban housing cheaper than other housing, but this was not primarily due to lowered real costs. Despite widespread application of mass-production methods in housing (Checkoway, 1980), productivity growth in house construction lagged behind manufacturing (Sims, 1980). Moreover, productivity increases applied more or less equally to all housing, so suburban houses enjoyed little disproportionate benefit. The suburban home's comparative advantage was not due to lower production costs.

The automobile, aided by federal highway policies, had more impact on cost, because low-cost transportation made inexpensive suburban land accessible. Decentralization lowered the relative price of central-city land and caused a massive redistribution of value from central cities to suburbs. This redistribution was inherently limited. It may have lowered the cost of living by lowering the cost of land, but suburban dwellings used more land per unit, so changes in cost per dwelling were smaller than changes in land costs.

Much of suburban housing's advantage came from federal subsidies. Financed by regressive tax measures, the system as a whole redistributed income from lower-income to upper-income households (Dolbeare, 1986). The subsidies therefore *raised* costs of living for the working class as a whole while channeling individual workers' consumption into suburban home ownership.

Suburban housing required high levels of individual consumption and was therefore costly in terms of both human and natural resources. Low-density settlements made shared consumption impractical, so each family unit purchased its own consumer durables. The suburban house's furnace took the place of the apartment building's boiler, the private laundry room substituted for the neighborhood laundry, and private kitchens preempted more collective food-preparation arrangements. Durable goods, many of which sat idle most of the time, were consumed in much higher numbers per capita than in other, less suburbanized, industrial countries (Groelinger, 1977).

Suburban living was itself labor intensive and used an enormous amount of human resources. It assumed each household had a full-time (female) homemaker and made less costly domestic alternatives, such as shared child care, difficult (Hayden, 1981). Millions of women were tied up doing housework instead of other activities. Similarly, suburbia brought with it many chores, such as lawn mowing, that were absent from other forms of housing. Instead of being done at the most efficient scale by skilled work groups using modern equipment, this work, including essential home maintenance and repair, was often done by unskilled, do-it-yourself home owners working individually. In short, postwar housing policies made suburban housing low in cost for individual families, but extremely expensive to society as a whole.

SOCIAL SEGMENTATION

The postwar housing system delimited housing tenure, location, and mobility options in ways coinciding with class position. Housing institutions amplified labor market segmentation by establishing spatially separate housing submarkets based on income, job stability, and other criteria that also differentiated labor markets. Overt discrimination further reinforced gender and racial differences in both labor and housing markets. Consequently, suburbanization primarily involved those incorporated into the mainstream of the political economy, while lower-quality, multifamily, central-city housing "trickled down" to those who were not.

Housing added new social differences. It accelerated the demise of the extended family by making it difficult for the elderly, families with children, and unmarried individuals to live near each other. Housing also divided persons with similar standing in the labor market along racial and ethnic lines. As a result, separate housing submarkets served different subsets of the population, and housing added new lines of social cleavage to those of the labor market.

HOUSING'S CONTRIBUTION
TO THE CRISIS

By the mid-1960s, signs of stress began to emerge. Inflation and unemployment soared while real earnings, productivity growth, and corporate profits plummeted. These changes were closely connected to social and technological limitations in postwar political economic arrangements. By the mid-1960s, most of the major refinements in assembly-line production methods had been accomplished, and further productivity increases became hard to achieve (Lipietz, 1986). Reforms in labor-management relations, unemployment insurance, and welfare had achieved social stability, but they also lessened the threat of job loss as an effective sanction in the hands of employers (Bowles, Gordon, & Weisskopf, 1984). The war in Southeast Asia tightened labor markets as well, further lowering the costs of job loss. Together, these forces caused productivity growth to stagnate. Committed to military intervention and faced with strong antiwar and social movements at home, Washington turned to debt financing rather than fan the flames of domestic discontent. Housing played an important role in the deepening crisis, and the housing system became increasingly problematic.

SATURATED MARKETS

By the late 1960s, the postwar U.S. city had been built, and the markets it had opened up were becoming saturated (Feldman & Florida, 1988). In the absence of wholly new models for consumption, expanded output could be absorbed only by incorporating peripheral sectors of the population, by public spending, by new markets, by systematic waste, or by demographic growth. Alignments in U.S. politics precluded the first two alternatives, and foreign competition precluded the third and fourth. Demographic growth was slowing, and the crisis slowed it further. As an alternative to expanding output, output was cut back and working hours reduced. If this had been done systematically, with no loss of disposable income, output could have been absorbed. Instead, hours were reduced through unemployment and the crisis deepened.

THE PRODUCTIVITY SLOWDOWN
AND THE PROFIT SQUEEZE

The housing system lowered the costs of job loss and thereby contributed to the overall productivity slowdown. Mortgage lending and short-term credit facilitated consumption at the same time bankruptcy legislation made foreclosure and repossession difficult. These arrangements had been implemented to facilitate consumer purchases and stabilize effective demand, but they allowed workers to put off repayments and to suffer income losses without an immediate decline in living standards. Consumer durables were long lasting, and unemployed workers could readily delay such purchases. Consequently, unemployment's bite became less painful.

Consumer expenditures did not grow as fast as wages and stayed in specific sectors, particularly those tied to housing. Housing structured consumers' purchases around certain commodities for which per capita demand had stopped growing. Keynesianism, social programs implemented during the 1960s, and postwar political alliances made the state a focus for political contestation (O'Connor, 1973). As the cost of job loss declined, workers and community groups sought welfare, environmental, and safety legislation that constrained capital and thereby accelerated the crisis of profitability (Weisskopf, Bowles, & Gordon, 1983). Stagnating productivity growth, wage levels resistant to downward pressures, and growing state expenditures combined to squeeze profits. Unable to reduce wages or public expenditures significantly, corporations increased prices to restore profit levels. The end result was a cumulative cycle of inflation.

THE STATE AND THE URBAN CRISIS

Segmented housing and labor markets eventually gave rise to political and social unrest. Metropolitan political fragmentation, which had allowed inequality to grow without redistribution through local government, now allowed disenfranchised groups to become influential forces in central-city politics. The smooth functioning of metropolitan areas was threatened, and national attention shifted to the inner cities. Federal urban programs grew enormously. Initially attempting to bolster corporate profits and restore social stability by incorporating peripheral groups (Davis, 1986), urban policy began to emphasize redistribution over physical renewal.

Yet powerful forces confined state action largely to providing public services, promoting public employment, and guaranteeing formal legal equality. Dominant economic ideology deemed public intervention legitimate only for so-called public goods, market "imperfections," and macroeconomic countercyclical measures. Entrenched political interests, representing multinational corporations, small business, and organized labor, opposed nationalization, direct public intervention, or massive redistribution. Consequently, federal programs fought poverty through such circuitous measures as health care (Medicaid) and preschools (Project Head Start). Given the segregation and fiscal-political fragmentation of urban areas, redistribution took place largely within the working class, often pitting Blacks against Whites. Thanks to urbanization and civil rights legislation, Blacks' economic status did improve, but during the worsening economic conditions of the 1970s this took the form of a growing bifurcation of the Black population: Some Blacks were incorporated into the social mainstream, but the rest were trapped as an urban underclass (Wilson, 1987). Because of these limits, urban programs were too anemic to stimulate the economy sufficiently to counteract the productivity slowdown.

THE COSTS OF SPRAWL

The "high standard of living" that had been the triumph of the U.S. economy suddenly became its major problem. Many of the high living costs in the United States were attributable to its settlement pattern. Urbanization is a physical, social, political, institutional, and economic process that cannot be altered at will, and, for this reason, living costs are difficult to change. When faced with permanent job loss, workers cannot manage to make sufficient wage concessions and still survive physically. Energy and other components of the unusually high costs in the United States were inherent in its urban form. In this way the housing system put a bottom limit on wages, and the crisis therefore took a variety of other forms.

Firms found it cheaper to move to distant locations and build entirely new plants than to rehabilitate old ones. Capital mobility and plant relocations only exacerbated the crisis (Bluestone & Harrison, 1982). Capital flight broke the production-consumption circuit. It also destroyed entire communities, leading to a series of localized crises, the effects of which spread throughout the entire economy. At the same time, mobile capital created strong spurts of growth elsewhere. Localized, often transient, booms caused severe strains and led some localities to overinvest in schools, roads, and other infrastructure.

THE EFFECTS OF
THE CRISIS ON HOUSING

While not the cause of the crisis, housing was interrelated with the crisis in several important ways. Low and uncertain profits in manufacturing made speculation in housing and real estate attractive. Coupled with the maturation of the baby-boom generation and a demographic shift to suburbs, smaller cities, and the Sunbelt, this led to feverish house price inflation in some markets while other markets stagnated. With households unable to afford the "American dream," new models for family living, such as dual-breadwinner families and shared quarters among unmarried adults, became increasingly common. These models ran afoul of other aspects of the postwar housing system and created new problems.

HOUSING FINANCE

As the crisis deepened, housing finance institutions experienced a growing tension. On the one hand, housing finance institutions were dedicated to home mortgage lending. On the other hand, the crisis made financial markets increasingly volatile. Enormous quantities of money were tied up in the housing sector. When funds became scarce in other capital markets, money could not readily get out of housing's separate financial market, and when funds were scarce in mortgage markets, capital from other sectors could not get in. Between 1965 and 1973, personal savings grew by 132%, from $30.3 billion to $70.3 billion, with deposits in savings accounts increasing from $28 billion to $64.2 billion. Over this same period, undistributed corporate profits declined by almost 17%, while corporate borrowing increased by 133%. Confined to housing, the overaccumulated funds leveraged a huge expansion of mortgage credit. Residential nonfarm mortgage debt shot up from $258 billion in 1965 to $550 billion in 1973, an increase of 113%. Only a small portion of this found its way into new construction: Between 1965 and 1973 the housing stock grew by approximately 22%. The rest went to refinance existing housing, leading to a dramatic rise in house prices.[1]

Severe and uneven fluctuations in house prices generated housing speculation and disinvestment side by side. Eventually this process undermined support for continuing to insulate the housing finance system from the rest of the economy, and bank deregulation followed. Since the housing finance system underpinned social and political organization, the crisis and reorganization of housing finance implied an overall restructuring of social patterns and political alignments.[2]

SPECULATION AND
HOUSE PRICE INFLATION

Housing also suffered from mounting speculation. As profitability in manufacturing and commercial sectors declined, investors sought out other investment outlets. Being a necessity tied to a nonrenewable resource, housing (and, more generally, real estate) has always been a prime candidate for speculation in times of falling profits. The high amount of leveraging and limited risk afforded by federal housing policy made housing speculation especially attractive. Between 1971 and 1977, two successive peaks in the housing cycle, mortgage debt on one- to four-family houses rose by over 350%, from $27 billion to over $95 billion. At the same time, capital investment in new plant and equipment declined from a high of 4% of GNP in 1966-70 to 3.1% in 1971-75 and 2.9% in 1976-80.

Responding to weak demand for business loans, banks joined savings and loans in the mortgage market (Grebler & Mittelbach, 1979, p. 104). Speculators invested heavily in housing to achieve profitability and protection from inflation. The result was massive and uneven inflation in house prices. Between 1968 and 1977, house prices more than doubled nationally, although regional and metropolitan price increases varied considerably (Grebler & Mittelbach, 1979). These increases far outpaced the overall rate of inflation: The consumer price index (CPI) increased by 75% and the GNP deflator, which includes price changes in intermediate goods, increased by 72%. Since these latter rates include a housing component, the difference between housing and nonhousing inflation was even greater than these figures indicate.

Some suggest that increased government regulation was the major culprit (President's Commission on Housing, 1982; Sternlieb & Hughes, 1980), but a close look shows this to be untrue. Price increases for *existing* homes, which were not heavily subject to increased government regulation, exceeded those of *new* homes throughout the period (Grebler & Mittelbach, 1979). Dowall (1984) estimates that government regulation added no more than 34% to new housing costs, and this was far below overall price increases.

Rapidly growing communities were a natural for speculation, and accelerated uneven development contributed to the general inflation. As profits stagnated in manufacturing, high rates of return on real estate became irresistible to investors. This put pressure on financial markets and resulted in their dramatic restructuring. Financial institutions developed a variety of devices for channeling capital into real estate. Real estate investment trusts (REITs), for example, permitted investors to speculate in real estate and share tax advantages with only limited exposure to risk. Large industrial corporations established their own real estate divisions: While Chrysler speculated in Arizona shopping centers, its auto plants in Detroit languished for lack of capital.

Areas that bore the brunt of economic contraction—most notably the industrial Midwest—saw housing prices appreciate at very low nominal rates or, in some instances, actually decline. Decreasing real home values exacerbated the effects of economic restructuring as declining home equity overlaid declining real earnings, further adding to the decline in living standards. Other areas, such as New York City and California, saw dramatic increases in house prices. Thanks to postwar financial arrangements, home buyers could easily leverage a 20% down payment into a better than 60% return on investment. Anxious to realize these kinds of returns, current home owners began using their built-up equity to "trade up" and to speculate in housing. Sellers commonly offered second mortgages to complete sales, thereby increasing leverage ratios. Multiple mortgages and balloon payments became common for the first time since the 1920s. Housing was thus transformed from merely providing shelter to being a hedge against inflation and a source of speculative windfalls (Sternlieb & Hughes, 1980).

HOUSING AFFORDABILITY AND THE CRISIS IN THE BUILT ENVIRONMENT

Spatial organization within metropolitan areas became increasingly problematic for growing numbers of people. With married women increasingly active in the labor force, especially in clerical, professional, and service sector jobs that tended to be centrally located, suburban locations lost much of their appeal. Suburban housing also had serious drawbacks for dual-breadwinner households (Van Allsburg, 1986). More important, the added income of two breadwinners, coupled with federal legislation requiring mortgage lenders to end their practice of discounting women's earnings, gave dual-breadwinner households considerable purchasing power. In turn, housing inflation and declining real wages created a crisis of affordability. House prices rose so high that few single-breadwinner households could afford to become first-time buyers. A feedback process had been set in motion whereby housing costs and female labor force participation increased hand in hand (Myers, 1985), as the growth in dual-breadwinner households counteracted declining real individual incomes.

THE CHANGING ROLE OF THE STATE

As the crisis developed, federal housing policy went through a series of corresponding changes. During the late 1960s, government responded to social unrest with an upsurge in housing assistance. A separate cabinet-level department, the Department of Housing and Urban Development (HUD), was established in 1965, replacing the lower-level Housing and Home Finance Agency. The Housing and Urban Development Act of 1968 increased federal aid for public housing and established programs for interest rate subsidies for low-income home ownership (Section 235) and rent subsidies to stimulate construction of multifamily rental housing (Section 236). Between 1969 and 1970 the share of all housing starts subsidized by the federal government more than doubled, jumping from 12% to 25% (Lilley, 1980).

Approximately as many public housing units were constructed in the 6-year period between 1968 and 1973 as were constructed over the 19-year period between 1949 and 1967, and operating subsidies for public housing increased over a hundredfold from 1969 to 1982 (Bratt, 1986).

In the 1970s, the focus of federal policy shifted from redistribution to flexibility and efficiency. The Housing and Community Development Act of 1974 combined a host of older categorical grant programs into lump-sum block grants. Community development funds were allocated by formula and had relatively few conditions attached to their use. These funds could be used to upgrade housing through rehabilitation, new construction, neighborhood redevelopment, code enforcement, self-help, and so on. Under Section 8 of the bill, public assistance for low-income housing shifted further from new construction toward rent supplements for new, existing, or renovated units.

The 1980s witnessed another major reorientation. The Carter administration had been unsuccessful at abating the crisis, and the Reagan administration employed a "cold bath" strategy of attacking the social wage while maintaining effective demand with military spending (Bowles et al., 1984; Struyk, Mayer, & Tuccillo, 1983). The administration adopted an explicit goal of minimizing federal involvement in housing. Specific objectives included completely privatizing federally supported mortgage markets (e.g., GNMA), substituting housing vouchers for new construction, tightening eligibility requirements, eliminating programs, at least partially privatizing the existing stock of public housing, and, most important, drastically reducing federal spending on housing.[3] Budget authority for subsidized housing, which peaked at $31.5 billion in 1978, fell to $13.3 billion in 1982 and $9.5 billion in 1987 (U.S. Congress, 1987, p. 8). Between 1981 and 1987, total federal housing budget authority, including subsidies, mortgage credit, and deposit insurance, declined from $29.3 billion to $16.3 billion (Executive Office of the President, Office of Management and Budget, 1982, 1988). As of 1988, HUD's budget had been slashed by more than two-thirds its 1981 amount and had fallen from fourth to eighth among twelve federal departments (Nenno, 1987). Nonfederal sources of new low-income housing construction were virtually eliminated by changes to federal tax laws in 1986; tax incentives, the major attraction for private investors, for constructing new low-income housing through either direct investment or tax-exempt Industrial Development Bonds were virtually eliminated. By the late 1980s, the federal government had essentially abandoned its commitment to housing assistance. New construction for public and assisted housing was virtually eliminated, and redistribution through federal housing programs was minimal.

The combination of declining housing assistance and a shrinking social wage weakened low-income groups' ability to obtain adequate shelter. At the same time, house prices skyrocketed in selected areas, leading to calls for "affordable housing." With little likelihood of assistance from Washington, state and local governments stepped in. Many states, even some with stagnant housing markets (such as Ohio), instituted programs to assist home buyers. Banks and other financial institutions also took up the cause. But the most notable feature of this movement was its social base. New Deal housing institutions had made suburban home ownership widespread among moderate-income Whites in traditional families, and precisely this constituency became the focus of "affordable housing" efforts. The clientele for government housing programs during the 1960s and 1970s—minorities, the elderly, and the poor—have been notably absent from these initiatives.

By the end of the 1980s the effects of restructuring had begun to be felt. Freed from prior regulatory constraints, hundreds of federally insured savings and loans had made speculative investments that went sour. Faced with impending collapse of the entire postwar financial system, the federal government stepped in with a historic savings and loan "bailout" in 1989.

Even though the housing system no longer functioned well for most of the population, the social polarization it fostered for over a generation now makes concerted political action highly improbable. The most likely result is even greater bifurcation of housing markets, further exacerbating the divergence of mass production from mass consumption.

SUMMARY

In response to a structural crisis, the New Deal completely revamped the U.S. housing system. In the immediate postwar period, housing's new institutional structure helped create unprecedented growth as mass production and mass consumption grew in parallel. But housing eventually became a barrier to further growth and a significant contributor to the crisis. Once housing-oriented consumption patterns were institutionalized and incorporated into social reproduction, they became resilient to change and acted as a constraint. The costs associated with suburbanization constituted the lion's share of the cost of living, and they were difficult to alter because they were rooted in particular social institutions and a distinct spacial form. Although the postwar commodity bundle was no longer growing fast enough, neither was it about to shrink. When the boom ended in the late 1960s, the consumption-intensive spatial form became increasingly dysfunctional. Housing and urban form put a lower bound on wages irrespective of labor's strength or weakness, and capital could not readily lower wages below this level without jeopardizing labor's continued reproduction.

Our argument may be summarized in five basic points. First, the problems in housing and mortgage lending did not emerge in isolation but were closely tied to more general economic problems: declining productivity and profit, rising inflation, tight labor markets, and burgeoning social unrest.

Second, these problems fueled an inflationary wage-price spiral that was uneven across sectors and especially pronounced in housing. At the same time, declining real earnings put strict limits on the ability to purchase housing, creating a scissor effect of rising prices and declining demand. Some households responded by increasing the share of income devoted to housing, but this was problematic because it shifted purchases away from other forms of consumption. The only households that even remotely kept up with galloping house price inflation were the growing numbers with two breadwinners. This too proved problematic because postwar housing was particularly ill suited to dual-breadwinner households. The once parallel development of mass production and mass consumption began to diverge.

Third, housing became a significant contributor to worsening economic problems. The scissor effect made it difficult for workers to purchase homes and put strict limits on housing's role as an investment outlet. Housing's ability to lead an economic expansion, as during the 1950s, was curtailed.

Fourth, housing inhibited economic restructuring. The high-cost housing and urban system could not readily respond to falling profits and intense international competition. The postwar built environment put a floor on living costs that translated into one on wages, and corporations could not lower labor costs without jeopardizing their work forces.

Fifth, various groups responded with different individual strategies that had important social, political, and economic implications. The most important strategy was the dual-breadwinner household. While the dual-breadwinner household propped up aggregate demand, the postwar suburban life-style was ill suited to the needs of dual-breadwinner households. The classic nuclear family received too little income to cope with rising costs of living, and those living in traditional families were forced to cut back on their consumption. This contributed to economic instability while simultaneously making the nuclear family increasingly problematic. The crisis was therefore more profound and fundamental than a simple crisis of profitability. It was simultaneously a crisis of profitability, social relations, social institutions, a form of urbanization, and a way of life.

The current housing situation in the United States may prove to be a social powder keg. Affordable middle-income housing is becoming scarce, and the prospects for low-income households are grim. As it ages, the baby-boom generation can be expected to put even more pressure on housing markets. In some areas, inflation has turned home ownership into a game of musical chairs in which some realize windfall gains, at least on paper, while the inflationary bubble does not burst. In others, many families are trapped as they watch their entire life savings being depleted through house price deflation. Given the highly uneven pattern of restructuring, the predicament of simultaneous inflation and devaluation is likely to worsen.

NOTES

1. Numbers cited in this paragraph are computed from the U.S. Bureau of the Census (1979, Tables 732, 733, 850, 1370, and 1376).

2. This is a very abbreviated summary of financial deregulation and the reasons for it. See Florida (1986) and Feldman and Florida (1988) for elaboration.

3. See Hartman (1986) for a review of these proposals and their impacts.

REFERENCES

Aglietta, M. (1979). *A theory of capitalist regulation: The U.S. experience* (D. Fernbach, Trans.). London: NLB.

Bluestone, B., & Harrison, B. (1982). *The deindustrialization of America.* New York: Basic Books.

Bowles, S., Gordon, D. M., & Weisskopf, T. E. (1984). *Beyond the waste land: A democratic alternative to economic decline.* Garden City, NY: Anchor.

Bratt, R. G. (1986). Public housing: The controversy and contribution. In R. G. Bratt, C. Hartman, & A. Meyerson (Eds.), *Critical perspectives on housing* (pp. 335-361). Philadelphia: Temple University Press.

Checkoway, B. (1980). Large builders, federal housing programmes, and postwar suburbanization. *International Journal of Urban and Regional Research, 4*(1), 21-45.

Davis, M. (1984). The political economy of late-imperial America. *New Left Review, 143*, 6-38.

Davis, M. (1986). The lesser evil? The Left and the Democratic party. *New Left Review, 155*, 1-36.

Dolbeare, C. (1986). How the income tax system subsidizes housing for the affluent. In R. G. Bratt, C. Hartman, & A. Meyerson (Eds.), *Critical perspectives on housing* (pp. 264-271). Philadelphia: Temple University Press.

Dowall, D. E. (1984). *The suburban squeeze: Land conversion and regulation in the San Francisco Bay Area.* Berkeley: University of California Press.

Executive Office of the President, Office of Management and Budget. (1982). *Budget of the United States government, fiscal year 1983* (97th Congress, 2nd session, H. Doc. 97-124). Washington, DC: Government Printing Office.

Executive Office of the President, Office of Management and Budget. (1988). *Budget of the United States government, fiscal year 1989* (100th Congress, 2nd session, H. Doc. 152). Washington, DC: Government Printing Office.

Feldman, M. M. A., & Florida, R. L. (1988, June 16-18). *Housing, Fordist decline, and economic restructuring in the United States.* Paper presented at the International Conference on Regulation Theory, Barcelona.

Florida, R. L. (1986). The political economy of financial deregulation and the reorganization of housing finance in the United States. *International Journal of Urban and Regional Research, 10*(2), 207-231.

Florida, R. L., & Feldman, M. M. A. (1988). Housing in U.S. Fordism. *International Journal of Urban and Regional Research, 12*(2), 187-210.

Grebler, L., & Mittelbach, F. G. (1979). *The inflation of house prices: Its extent, causes, and consequences.* Lexington, MA: Lexington.

Groelinger, D. (1977). Domestic capital equipment. In J. E. Ullman (Ed.), *The suburban economic network: Economic activity, resource use, and the great sprawl.* New York: Praeger.

Hartman, C. (1986). Housing policies under the Reagan administration. In R. G. Bratt, C. Hartman, & A. Meyerson (Eds.), *Critical perspectives on housing* (pp. 362-376). Philadelphia: Temple University Press.

Hayden, D. (1981). *The grand domestic revolution: A history of feminist designs for American homes, neighborhoods, and cities.* Cambridge: MIT Press.

Lilley, W., III. (1980). The homebuilders' lobby. In J. Pynoos, R. Schafer, & C. W. Hartman (Eds.), *Housing urban America* (2nd ed., pp. 32-50). New York: Aldine.

Lipietz, A. (1986). Behind the crisis: The exhaustion of a regime of accumulation. A "regulation school" perspective on some French empirical works. *Review of Radical Political Economics, 18*(1/2), 13-32.

Myers, D. (1985). Reliance upon wives' earnings for homeownership attainment: Caught between the locomotive and the caboose. *Journal of Planning Education and Research, 4*(3), 167-176.

Nenno, M. K. (1987). Reagan's '88 budget: Dismantling HUD. *Journal of Housing, 44*(3), 103-108.

O'Connor, J. (1973). *The fiscal crisis of the state.* New York: St. Martin's.

President's Commission on Housing. (1982). *Report of the President's Commission on Housing.* Washington, DC: Government Printing Office.

Sims, C. A. (1980). Efficiency in the construction industry. In J. Pynoos, R. Schafer, & C. W. Hartman (Eds.), *Housing urban America* (2nd ed., pp. 358-371). New York: Aldine.

Sternlieb, G., & Hughes, J. W. (1980). The post-shelter society. In G. Sternlieb, J. W. Hughes, R. W. Burchell, S. C. Casey, R. W. Lake, & D. Listokin (Eds.), *America's housing: Prospects and problems* (pp. 93-102). New Brunswick, NJ: Center for Urban Policy Research.

Struyk, R. J., Mayer, N., & Tuccillo, J. A. (1983). *Federal housing policy at President Reagan's midterm.* Washington, DC: Urban Institute.

U.S. Bureau of the Census. (1979). *Statistical abstract of the United States, 1979.* Washington, DC: Government Printing Office.

U.S. Congress, House Committee on Banking, Finance and Urban Affairs, Subcommittee on Housing and Community Development. (1987). *Hearings on the Housing, Community Development, and Homelessness Prevention Act of 1987, Part 1* (100th Congress, 1st session, March 11-12). Washington, DC: Government Printing Office.

Van Allsburg, C. M. (1986). Dual-earner housing needs. *Journal of Planning Literature, 1*(3), 388-399.

Weisskopf, T. E., Bowles, S., & Gordon, D. M. (1983). Hearts and minds: A social model of productivity growth. *Brookings Papers on Economic Activity, 2*, 381-441.

Wilson, W. J. (1987). *The truly disadvantaged: The inner city, the underclass, and public policy.* Chicago: University of Chicago Press.

3

Local Government, the State, and Housing Provision: Lessons from Australia

CHRIS PARIS

MANY SCHOLARS IN EUROPE and North America use the experience of their own countries to form the basis for wider generalizations about central-local relations. Gurr and King (1987) recently have summarized much of the diverse and varied literature on the autonomy of the "local state" during a period of fiscal stress. They suggest that growing central restraint on local governments in the United States and Britain may be indicative of a general reversal of the 30-year trend toward greater public spending on local service provision within advanced capitalist societies. They suggest that "a substantial restructuring of state-city relations is likely to follow as the local state is pushed into assuming greater responsibility for financing local services" (p. 72).

Such a generalization does not hold for Australia, where there is virtually no metropolitan government. The commonwealth government has not played a major role in reforming urban governments; rather, "the most significant impulse for reforms has emerged at the State level" (Power & Halligan, 1987, p. 85). The Australian case, therefore, has some important implications for general theories about the state-city relations in Western societies. It is sufficiently different in enough ways from other advanced countries to constitute an important test of

AUTHOR'S NOTE: An earlier version of this chapter was presented at the 1988 International Conference on Housing Policy and Urban Innovation, Amsterdam, June 27-July 1, 1988. This chapter has been improved by comments from John Halligan and Marian Simms.

generalizations developed elsewhere. In particular, debates concerning the "local state," the dual-state thesis, and housing class all look idiosyncratic from an Australian perspective. Australia is distinctive in terms of (a) its political system (federalism, with the states and territories being a more powerful form of noncentral government than is local government), (b) its economic and social history (settler capitalism, little industrial development, substantial immigration, and postwar affluence), and (c) its unusual history of urban development (massive metropolitan dominance, low-density suburban sprawl, rapid development of tertiary and quaternary urban forms).

I have argued elsewhere that the concept of "the local state" is flawed theoretically and, from an Australian perspective, appears to be little more than a slogan providing apparently theoretical status for local government (Paris, 1983). Recent developments concerning local government and housing in Australia reinforce my contention that subnational units of government differ sufficiently between countries and over time to render the concept of the local state inoperable as a category for comparative historical analysis. The institutional structure of the Australian state has crucially influenced the capacity for change in the range of functions situated at different "levels" of government as well as limited the scope for the transfer of housing functions to Australian local government.

This chapter explores some of these issues in Australia. The distinctive structure of housing provision is explored, together with some of the changing relationships between governments and housing, emphasizing the limited role of local government generally and specifically regarding housing policy. The concluding section considers the significance of recent developments in Australia for international debates.

HOUSING PROVISION IN AUSTRALIA

The combination of factors affecting access to housing has varied both over time and among places in postwar Australia. How best can we conceptualize that specific national housing history as well as relate it to similar housing experiences elsewhere? A good starting point is Ball's (1983) notion of structures of housing provision: "Housing via a specific tenure form is the product of particular, historically determined social relations associated with the physical processes of land development, building production, the transfer of the completed dwelling to its final user and its subsequent use" (p. 17).

This idea correctly emphasizes the need to see particular tenures as historically distinctive social relationships and not as ahistorical and universal ways of demarcating rights in residential property. The idea of structures of housing provision can be usefully extended beyond Ball's tenure-specific focus, however, to encompass overall patterns of finance, construction, distribution, and access throughout a housing system. Different institutional forces affect access to different tenures in different ways, but in particular places at particular times there are definite combinations of housing construction, existing dwellings, institutions (public and private), and households. Overall structures of provision both set limitations on the kinds of housing opportunities to which people can aspire and create patterns of opportunity within which individuals and families can exercise choice.

The structure of housing provision in Australia has resembled that in the United States rather than that in Britain or other European countries. This has reflected both Australia's more "privatized" cities (Kemeny, 1981, 1983) and forms of governance (Parkin, 1982). Australian housing has been produced within a distinctive history of colonial settlement and subsequent "settler capitalism" (Denoon, 1983). Separate colonies were not united into a nation until federation, in 1901. Boom/bust economic development in the nineteenth century had been dominated by pastoralism and periodic mineral booms. There was only limited development of local manufacturing industries. Urban and regional development resulted in metropolitan dominance, with the bulk of the population living in a few coastal centers (Logan, Whitelaw, & Mackay, 1981). Reviewing Australian housing issues and policies, Kendig and Paris (1987) claim: "The Australian housing system today is distinctive, compared to European and North American countries, both in physical terms and in its tenure composition. The housing stock is generally newer and . . . at very low densities. The level of home ownership is very high and . . . public and social rental housing extremely low. The private rental sector has been sizeable" (p. 9).

The development of the Australian tenure system (see Figure 3.1) has reflected both the changing economic fortunes and government policies and priorities. Growth in the housing stock after 1945 was fueled by a postwar baby boom as well as by growing in-migration during a long period of prosperity. The postwar economic boom and favorable commonwealth government policies enabled substantial growth in the level of home ownership, especially during the 1950s. The private rental sector boomed during the 1960s, as rapid growth in school leavers and continued economic prosperity resulted in high levels of effective demand for rental housing. Since the early 1970s, however, there has been growing polarization of housing options, a

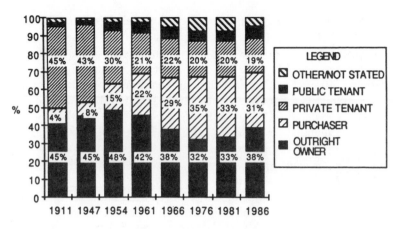

Figure 3.1. Private Households by Tenure, Australia, 1911-86

cessation of the growth of home ownership, and the emergence of new housing problems: growing waiting lists for public housing, substantial youth homelessness, and a major problem of affordability facing poor private tenants. Even so, most Australians are extremely well housed in relatively new dwelling stock built at the lowest suburban densities in the world.

GOVERNMENTS AND HOUSING
IN AUSTRALIA

Relationships between governments and housing are entangled within the complexities of the Australian federal system. Figure 3.2 indicates some of the relationships among the three "levels" of Australian government and their housing policies. The study of federal societies once used a "layer-cake" metaphor to discuss the various levels of government, as if there were clear lines of demarcation and directions of power and influence between them. In Australia, as elsewhere, that metaphor has long been replaced with one of a "marble cake," within which there are huge areas of complex interaction and a systematic blurring of the differences among levels of government. The commonwealth and the six main state governments are based in the Constitution, whereas local government systems are the separate creatures of the individual state legislatures.

Any analysis of local housing issues needs to recognize the limited role of local government within Australia. The history and structure

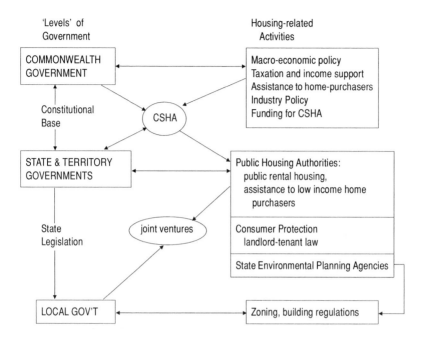

Figure 3.2. Levels of Government and Housing Activities in Australia

of Australian federalism has resulted in intergovernmental relations within which local government is fragmented, small, poor, and weak. Local government accounts for only about 6% of total public expenditure, compared with the 20-25% spent by local governments in other federal countries, such as the United States, Canada, and Switzerland. Australian local government derives about half of its income from property taxes (rates) and has focused its activities primarily on a narrow range of functions related to the service of property: local roads, rubbish collection, and the limited provision of water, sewerage, and other infrastructure items. Local roads usually constitute between one-third and one-half of councils' expenditures. Major functions provided by local government in other countries, such as education, social services, emergency services (police, fire prevention, and hospitals), and housing, are primarily administered by state rather than local government.

Both commonwealth and state governments allocate funds to various housing policies as well as influencing housing through economic and other policies (for a comprehensive recent review, see Kendig & Paris, 1987; see also Flood & Yates, 1987). Local government expenditure on housing is confined to small, localized programs. Most local government direct involvement in housing is through zoning and other land-use and building regulatory activity.

The commonwealth government affects housing directly and indirectly through macroeconomic and fiscal policies, taxation, income support, and industry policies as well as housing policy. Its specific housing policies and housing-related policies are dispersed among numerous ministers and their departments. The most visible housing policies currently are the responsibility of the Department of Community Services and Health, including direct assistance to home purchasers and the provision of financial assistance to the state housing authorities (currently in the form of grants) allocated through periodic joint agreements between the commonwealth and the states: the Commonwealth-State Housing Agreement (CSHA). CSHAs since 1945 have funded state rental housing provision and assistance to low-income home purchasers. However, the 1984 CSHA has encompassed a wider range of policy initiatives, including the Local Government and Community Housing Program (LGACHP), which encourages local government and community housing initiatives.

Western Australia, Victoria, South Australia, and New South Wales all had Australian Labor Party (ALP) governments between 1984 and March 1988. New South Wales was won by the conservative Liberal-National Coalition, however, in March 1988. With an ALP government in Canberra since early 1983, Labor had a majority at CSHA negotiations between 1983 and 1988. All states have great capacity to frustrate commonwealth initiatives through the relatively strong position of the states within the Australian Constitution, but the Labor majority at CSHA negotiations produced greater coherence between commonwealth and state housing policies after 1983 than during the preceding decade.

Housing authorities in the Labor states have moved toward a comprehensive approach to their housing roles during the 1980s, although their expenditure is still dominated by the two established program areas of public rental provision and assistance to home buyers. Labor governments in Victoria and New South Wales wanted to have greater involvement in housing policy. They abolished the former statutory authorities, which had some autonomy from the governments of the day, and instead incorporated them into departments with direct ministerial control. Most Labor states also moved toward a more overtly politicized bureaucracy with the appointment of senior officials who are sympathetic to the ALP. Classic models of a "neutral" public service, which often used in any case to obscure the political views of senior officials who strongly opposed ALP priorities, are difficult to sustain, as both Labor and non-Labor governments replace senior public officials after changes of office.

State government housing authorities, over the past 20 years, increasingly have housed the poorest Australians, performing an ever more limited "welfare" role rather than moving toward wider social housing provision. Public housing has become a form of de facto income assistance selectively available only to public tenants. The increasing proportion of poor public tenants has led to a situation in which (rebated) rents are not able to meet recurrent costs, and state governments are facing a mounting fiscal crisis in their rental operations. Meanwhile, poor private tenants receive virtually no assistance apart from a small additional supplement to pensioners and other welfare beneficiaries. The states also provide assistance to low-income households for home purchase. This home purchase assistance has been targeted toward poorer households much more successfully than has been the case with the various schemes administered directly by the commonwealth (Flood & Yates, 1987). State governments, like the commonwealth, pursue many other housing programs, but these account for relatively small proportions of housing expenditure. The Labor states generally have encouraged and supported a variety of local government and community housing initiatives, with New South Wales being most active in this area. State governments are also responsible for consumer protection (including landlord-tenant law) and environmental planning.

The commonwealth has few direct links with local governments, which are the creatures of state and territory legislation. Commonwealth responsibility for local government is located in the Office of Local Government (OLG), under a minister for local government who is a junior beneath the minister for immigration, local government and ethnic affairs. Local government has played virtually no role in direct housing provision except in isolated rural communities where it is difficult to attract qualified staff. Few councils have provided other rental housing except for special groups (e.g., the frail elderly). Local government's main impacts on housing have come through zoning and building regulation policies and practices. Other local government activities, particularly regarding local roads and environmental works, have affected housing indirectly.

There was growing interest in enlarging the housing role of local government during the 1980s. Local government and community organizations became more active in direct housing provision and through joint ventures with state governments. Local communities, through local government authorities and voluntary organizations, sought additional resources from state and commonwealth governments to enable them to play new roles in housing provision. Politicians and bureaucrats within commonwealth and state governments, however, have tried to encourage more local generation of resources to be allocated to the solution of local housing problems.

LOCAL-LEVEL INITIATIVES

The traditional housing role of local government has come under increasing pressure for reform since the late 1970s. There has been a groundswell of activism and enthusiasm in many communities. The main impetus for a new localism, however, has come from "higher" levels of government. State governments, particularly in New South Wales, encouraged and assisted local initiatives. Such developments were institutionalized nationally within the 1984 CSHA in the Local Government and Community Housing Program. The Commonwealth Office of Local Government undertook a series of local housing projects, especially studies of local housing markets. The process of innovation with regard to local government and housing overall, therefore, has been more "top-down" than "bottom-up."

One element in the renewed interest in local government and housing has been the growth of a "human services" perspective within local government, especially in the states of Victoria and New South Wales. This has been associated with increased politicization of local government, though there is still a strong belief that local government should be "nonpolitical" (Halligan & Paris, 1984).

The early 1980s were years of housing reform. One strand within state government thinking was that local government had generally been too conservative, both through ignoring the effects of various council programs on housing (especially zoning) and through deliberately operating policies of exclusionary zoning. The encouragement of community groups and "progressive" elements in local government was intended both to stimulate local government response and to be part of a changing, more "community-oriented" public housing role. The increasingly politicized bureaucracies also took on many junior staff, often on temporary contracts, who previously had been community activists. Many former activists were recruited on the basis of their experience and enthusiasm, thus reducing the number of community-based housing activists! At times, indeed, the views of state bureaucrats were indistinguishable from those of housing activists outside. The March 1988 state election in New South Wales resulted in defeat for the ALP. The new Liberal-National Coalition removed the director of housing and other senior housing officials. Housing policies have been reviewed, and there has been substantial winding down of many Labor initiatives. Community-based programs have been abolished, tenants' participation programs have been defunded, public housing is scheduled to be sold, and a more "market-oriented" approach is being taken to state housing policies.

There was less widespread politicization of the commonwealth bureaucracy overall since the 1983 election of the Hawke Labor government, though this process has been in evidence. Some individuals within government have played crucial roles in forming the commonwealth government's approach to local government and housing. Tom Uren, the minister for urban and regional development from 1972 to 1975, was made minister for territories and local government shortly after the Hawke Labor government was elected in early 1983. He had been a leading member of the parliamentary left for many years and was a strong proponent of decentralization and local autonomy. He had been opposition spokesperson on housing and an advocate of major reforms in housing policy. As a minister under Hawke, Uren strongly lobbied for a more positive commonwealth role in local government, and initiated an inquiry into local government finance. Apart from Uren and his staff, however, support for local government housing initiatives has never had a very high priority within the Hawke government. Varying views concerning commonwealth involvement in local housing are held both within the Parliamentary Labor Caucus and within various commonwealth departments. The "dry" view, which has considerable weight within cabinet and caucus, is that the commonwealth should withdraw from, rather than increase its involvement in, such activities. Left Labor parliamentarians, however, want the commonwealth actively to support and encourage local initiatives.

The next two sections of this chapter focus on the two main commonwealth initiatives regarding local government and housing. These programs illustrate two aspects of the general problem facing the commonwealth in any attempt to become more closely involved in local government housing issues. The first explores some of the complexities of trying to relate local housing issues to wider housing systems and processes of change. The second reviews the constraints on any more positive roles for the commonwealth in encouraging or facilitating local action and for local government authorities that want to widen the scope of their housing action. Both cases reveal the ways in which the structure of Australian federalism renders simple models of central-local relations extremely problematic.

PROBLEMS OF RESEARCH:
THE LOCAL HOUSING MARKET STUDIES

The Office of Local Government, established in 1983, has been the principal commonwealth government agency dealing directly with local government and local government organizations. One of its main programs has been the Local Government Development Program (LGDP), which has sponsored, among other things, research into and development of the role of local government in economic development. A series of housing projects were funded within the LGDP "to raise the awareness of local authorities as to their direct and indirect impacts on the local housing market." These housing projects were also intended to explore how the supply of low-cost rental housing could be increased. The housing projects derived in part from the strong commitment of politically active senior officers in the OLG, one of whom previously had been a senior ALP staffer for Minister Uren.

One objective of the OLG housing studies was for councils located in areas of high housing priority "to analyse the structure and dynamics of their local housing stock and markets." Most studies were done by consultants working on briefs from councils or regional groups of councils, in consultation with the OLG. Studies were funded in most states and in a variety of locations, ranging from inner metropolitan through suburban metropolitan and commuter-zone areas to a small isolated town. Some of these projects took regional approaches to groups of councils comprising both outer metropolitan and nonmetropolitan councils. The councils varied considerably in terms of economic activities, population, and area.

Little guidance was given to the consultants undertaking the various housing market studies regarding the conceptual basis of the terms *local housing markets* or *dynamics* of housing. The consultants effectively interpreted the term *local housing market* as equating to the housing stock and households within the geographical jurisdiction of the authorities being studied. Little consideration was given, however, to the extent to which these spatial units represented distinct or discrete housing markets or submarkets. None of the studies addressed the question of what might be meant by the *dynamics* of a local housing market; rather, they focused much more narrowly on trends in demand for housing and housing supply. Little attention was paid to processes affecting housing access.

The studies varied widely in terms of their objectives, methodologies, and presentation. Even the most sophisticated and professional studies, however, paid scant regard to wider national factors affecting local housing. Factors such as changing levels and patterns of international migration, the deregulation of the financial system and the weakening of the Australian dollar, taxation reform and changes to income support measures, however, all were having, or were likely to have, systemic effects on housing.

State housing and other policies have operated, in combination with market processes, to create distinctive differences among states, particularly concerning the size of the public rental sector. Regional economic performance, too, has affected housing crucially. Country districts affected by the relatively poor returns on agricultural investment have experienced depressed house prices. Rapid economic growth that has stimulated in-migration in many coastal areas, in contrast, has rapidly raised rents and boosted house prices. Housing costs and availability vary substantially both among and within different geographical areas.

Such differences reflect the interaction of factors influencing housing, operating at different spatial scales, factors that are not themselves entirely separate and discrete. The various combinations of dwellings, households, tenure, costs, and so on, therefore, have not coalesced into discrete and homogeneous "local housing markets". Most of all, given the fragmented nature of most of local government, spatial variations within metropolitan areas have not produced distinctive submarkets that coincide with municipal boundaries.

These arguments are not intended to deny that some council policies have very important effects on housing both within their own local government areas and in adjacent authorities. Zoning policies regarding flat construction, density standards, and types of permissible dwelling all influence the nature of housing within such authorities. Such variations help to explain the dramatic differences in terms of the size of the private rental sector and the nature of the built environment between those councils that allowed flat block development in the 1960s and 1970s and those that did not. Building regulations, also, have affected the financial viability of many projects. The speed with which planning applications are processed, especially during periods of high holding costs, has influenced what has been built, where, and by whom. The planning and building control policies frequently have spilled over into adjacent areas.

The point of emphasizing the limited value in conceiving local housing markets in terms simply of housing within municipal boundaries is that different combinations of the many dimensions of overall housing systems occur within different local authority areas. Some distinctive "types" of local authority housing systems, however, can be identified in terms of regularly recurring combinations of factors. Very little work has been done on this in Australia, but some preliminary analysis of census data has been done in house relating to local authorities in New South Wales. On the basis of that preliminary analysis of census data and with the addition of one category not found in New South Wales, Table 3.1 contains a schematic typology of local authorities in terms of the combinations of elements of the overall structure of housing provision.

The combinations of elements that make up "local" housing systems thus correspond to inner/middle/outer suburban metropolitan typologies and to a rural/resort/coastal typology of nonmetropolitan areas. There are other dimensions of variation, particularly in terms of house prices, rent levels, and intensity of competition among users. Most local authorities, however, can be fit broadly within the range of types set out in Table 3.1: inner city, metro-suburban, outer metropolitan, commuter area, country town, rural, resort, and mining center.

Few local authorities, of course, perfectly fit the different types in this classification; rather, such a typology provides a first approximation for identifying councils. Some contain a mixture of two different types. For example, some coastal local authorities contain both high levels of retirement and resort activities, producing a blend of the characteristics of established metropolitan areas (high levels of outright home ownership among elderly households) and seasonal resort rental variations. By contrast, some very small authorities within large metropolitan areas can have very high concentrations of only one or two elements of metropolitan housing systems overall. Some variations among cities, moreover, reflect their relative ages and the main periods of their growth.

One of the most valuable outcomes of the OLG housing studies has been the clear demonstration of the futility of trying to equate separate "local housing markets" with the territorial jurisdictions of local government authorities. There are regularly occurring patterns of housing factors in different kinds of local government authority. This, however, only emphasizes that local government areas or even "national" or

TABLE 3.1

A Typology of Australian Local Authorities in Housing Terms

(1) inner city	relatively low level of homeownership, very high level of private rental, high level of public rental; generally high but variable house prices, very high rents; oldest housing stock, some redevelopment and most gentrification
(2) metro-suburban	average or above-average levels of outright owner occupation, average to low level of purchasing homeownership, high level of private rental; highly variable house prices, average to high rents; mixed ages of housing, some gentrification
(3) outer metropolitan	low levels of outright ownership, high levels of mortgaged ownership, low levels of private rental, some very high levels of public rental; varying housing costs for owners and tenants, but generally lower than 1 and 2; newer housing
(4) commuter areas	average to below-average levels of outright owner occupation, average to high levels of mortgaged homeownership, low levels of both private and public rental housing; substantial cost variation; both "inland" and coastal examples; new and older housing stock
(5) country town	above-average levels of outright ownership, below-average levels of mortgaged homeownership, below-average level of private rental; highly variable levels of public housing; generally low house prices and rents, rare "town-camping"; usually older housing stock
(6) rural	very high level of outright owner occupation, low levels of all other tenures; very low housing prices, some "town-camping"; usually older housing stock
(7) resort	high levels of outright homwownership, low levels of mort-gaged ownership, high levels of private rental housing, little public housing; average to above-average house prices, large seasonal variations in rents; mainly new housing, including commercial retirement development
(8) mining town	low levels of owner occupation, high levels of rental and company housing, relatively transient and unstable housing markets; often very high housing costs; new and often mobile housing stock

SOURCE: In-house unpublished consensus analysis.

"state" housing systems are only optional levels of aggregation among a range of levels available for analysis. There can be no general abstract resolution to the question of what constitutes the most appropriate level of aggregation—it all depends on the questions that are being asked. In terms of the factors influencing change within local government areas, local data must be analyzed within the context of overall metropolitan processes rather than focusing within the relatively accidental history of local government boundaries.

The OLG studies also have thrown into relief the importance of assessing councils' existing policies and practices and their effects on housing, especially zoning and building regulations. The studies have also helped the councils involved to think through the scope for possible council roles in the future, and they have provided the basis for much better informed discussion of the scope of local government housing action. Finally, these research studies have helped to emphasize the diversity of housing circumstances within the 800-odd local government authorities that would confront any greater commonwealth involvement. The prospect of devolving commonwealth housing functions to that fragmented and fiscally weak local government system, even if it were possible to bypass the states and territories, is daunting.

PROBLEMS OF ACTION: THE LOCAL GOVERNMENT AND COMMUNITY HOUSING PROGRAM

The Local Government and Community Housing Program was established within the 1984 CSHA. It has been a "mixed" program, bringing together two previously separate strands of community housing initiatives and local government housing initiatives. It had been proposed to operate the two strands of the program separately, but the level of funds allocated would have resulted in greater administrative expenditure on what has been an expensively administered and inefficient program. LGACHP has been allocated modest funding, reflecting its low priority within direct commonwealth housing expenditure ($11 million in 1986-87 and $12 million in 1987-88). Some senior commonwealth housing officials strongly support locally oriented initiatives, but others see the LGACHP as little more than unnecessary window dressing or as a placebo for the left.

The two strands within the LGACHP have sat uneasily together. The commonwealth government grouped together local government and the voluntary/community sector rather than defining local government as a partner with the commonwealth and states in a democratic system of elected government. Many of the commonwealth bureaucrats had far closer links with and greater sympathy for the community sector than they did for local government.

The commonwealth did not devolve responsibility for LGACHP to the states because commonwealth officials themselves wanted to retain control of the program. Just as commonwealth and state governments are reluctant to hand over control to local government, so too the commonwealth rarely devolves powers to the states. Despite low priority overall, the LGACHP was given high priority by those commonwealth bureaucrats responsible for its implementation because they wanted others to see it as an important program. The small amount of money available under this program, moreover, has been sufficient to encourage many housing activists to concentrate on the art of grants-personship rather than to focus on the overall directions of commonwealth policies regarding housing.

The hidden agenda of many commonwealth bureaucrats for the LGACHP has been based on the low esteem in which they hold local government and their view that local government should provide more resources for housing. One stated objective of the LGACHP, therefore, has been to attract resources that otherwise would not have been used for housing investment or provision, for example, to tap potential sources of local finance—from both local governments and local communities—or to utilize council-owned land. Virtually no evidence has been made public with which to evaluate whether this objective is being met, although officers involved in the program have claimed that this was one area in which the policy was working (Harnisch, Harmer & Moore, 1986). However, they examined neither whether other, cheaper policy options could have achieved the same objectives nor whether such funds or resources might not in any case have been made available.

Most of the projects funded during the first few years of LGACHP did little more than duplicate existing housing programs, especially in the provision of aged persons' accommodation. Many elderly people are in housing need, but most are better off than many other groups. Needy elderly households, moreover, have had high priority within well-established state housing programs. The LGACHP also has funded provision for youth housing—an important area of emerging need—but virtually identical projects have been funded under another CSHA program—the Crisis Accommodation Program.

LGACHP has not merely duplicated existing provision, but may have done so at greater expense than existing modes of provision. It is difficult to judge the effectiveness of the program in cost-benefit terms, however, because data collection and publication so far have been poor. There has been no public evidence of monitoring or usable public records of expenditure. This makes it extremely difficult to measure the program against other possible courses of action. Unless the common-wealth does have adequate data, the LGACHP remains a "hopeful signal" that is incapable of being effectively monitored (Hogwood & Gunn, 1984, p. 222).

There have been major operational differences between the states. Also, the LGACHP is essentially running against the supposed thrust of the 1984 CSHA. Rather than directing resources to those in greatest need, the program has funded projects on an ad hoc basis, supporting bids as and when they came in, without any regular public accountabil-ity. The program has stimulated interest and supported some innovative activities, but its overall effect on housing provision has been slight.

The combined effect of commonwealth, state, and some local gov-ernment initiatives has done little to modify the traditional housing role of local government. Some councils have been made more aware of the impacts of planning and other policies on local housing, although the findings of the local housing studies were not widely disseminated. There has been growth in joint venture activities, in which, for exam-ple, a council provides land on which the state housing authority con-structs rental dwellings. Some councils have appointed or designated officers with specific housing responsibilities. Councils have become involved a little more in housing issues, but the response has been extremely uneven throughout Australia. There has been some growth of employment related to these developments, especially within state housing authorities. The community/voluntary housing sector has prob-ably grown more significantly than local government.

LOCALISM: SYMBOLIC POLITICS
OR STRATEGIC WITHDRAWAL
BY CENTRAL GOVERNMENT?

One of the key scholarly debates about state housing (and other) policies during recent years has concerned the changing relationship between central and noncentral elements of modern states. This has been sharpened as conservative national governments have sought to reduce public expenditure while relegating some functions from central to local responsibility. The growing interest within the Australian com-monwealth government for more local government involvement in housing could be taken to be indicative of such a trend. Has the support for localism been a smoke screen facilitating strategic withdrawal by the commonwealth from housing intervention? What, if any, have been the effects of the commonwealth research and action initiatives on local government and housing? Has there been any movement away from central state expenditure on housing and transfer of responsibilities for social reproduction to localities?

The answers to these questions are inevitably complicated by the intergovernmental relations of Australian federalism. The CSHA is a negotiated agreement whereby the states retain a strong role. Australian local government is not party to the CSHA. Local government, moreover, is fragmented among and within states. The great diversity of local housing circumstances revealed by the OLG studies reinforces the difficulty of making needs-based allocations without major reform, including amalgamation, of local government. Despite local government pretensions about being "nearest to the people," there are very low turnouts for local government elections. Councils have been dismissed with little popular protest. Strong opposition to amalgamations reflect the views of a well-organized minority rather than mass support of the existing local government system.

Commonwealth support for localism, in practice, has been limited. The commonwealth has allocated a tiny proportion of its housing expenditure to the LGACHP, and the funding of the local housing projects within the OLG was modest and has been discontinued. Crucially, neither the commonwealth nor the states have sought to compel local government to do more. The commonwealth, indeed, lacks the powers to compel local government to change its role; rather, it must seek to persuade, tempt, or shame local government into a more active role.

The main problem for the states has been growing inability to fund their activities from rents paid by their tenants. State government expenditures on housing have increased significantly, but the states have been unable to shorten waiting lists due to the commonwealth's refusal to count rent rebates as income support expenditure. Global housing funds have to be used to pay for rebates rather than an expansion of the stock.

Commonwealth housing policies have continued to give highest priority to the stimulation of construction and the support of home ownership. The recent growth of housing problems has been met with renewed activity to assist home purchasers rather than the expansion of other forms of provision. The group in most housing need—poor private tenants—has been neglected.

Much opposition has remained in local government to a wider housing role. Some have argued that it simply is not the business of local government to widen its range of functions. Given its limited resource base, local government has no capacity for undertaking anything but the most modest income support or redistributive activities. Elected representatives, probably correctly, anticipate a major ratepayers' backlash in response to "unnecessary" expenditure. Local governments have been more active in regard to local economic development than to human services provision. Whereas local economic initiatives have the potential to increase local government revenues, through an expansion of the rate base or by using council resources to generate additional income, human services provision typically increases local expenditure without any corresponding increase in income.

As yet, therefore, there is no evidence to support the notion that there has been any shifting of the burden of social welfare dimensions of housing policy from the commonwealth through the states onto local governments. The structure of the Australian state may actually make it more difficult for such a process to operate than in unitary states. A widened local government housing role would require prior far-reaching reform of commonwealth income support and other housing-related policies. The states, however, already have picked up a greater share of public expenditure on housing. If there were to be any concerted attempt to force local government to do more, it would have to come from the states rather than from the commonwealth. The weak and fragmented nature of local government means that it would have to be subject to major reform before it could have other functions imposed upon it. There has been some movement toward regionalism. This has involved both the devolution of some state management functions to regional offices and also the liaison between groups of councils to provide some services collectively. Such early initiatives carry the possibility for more far-reaching change, but as yet they have not had any systemic impact. Local government reform would have to involve amalgamation of existing councils, greater local autonomy, and greater scope for locally generated revenues. Such changes, moreover, would have to occur on a state-by-state basis before anything resembling a new national system could be created. Such reforms, while possible, are not on current political agendas. A more likely scenario, particularly in the light of continued local government resistance to reform, is a gradual decline in the significance of local government within the Australian body politic.

The commonwealth government, in conclusion, has not tried to stimulate new local government housing roles as part of any strategic withdrawal from its housing functions. Advocates of an expanded local government role in the OLG were trying to enhance the status of and establish a broader base for local government. Supporters of the

LGACHP within the commonwealth bureaucracy, too, have tried to expand the program's resource base from CSHA funds. The effect of the local housing studies has been to provide a basis for improving our understanding of the interrelations between "local" and larger-scale housing systems, though these studies have not been well publicized and most local councils probably are unaware of their existence. The studies, as well as LGACHP, have also underlined the very limited capacity for a wider, more positive local government housing role unless there are major institutional and policy reforms.

Any major strategic reorganization of the state overall is more likely to involve changes in the relationship between the commonwealth and the states rather than any "lower" levels of government. The Australian case discussed here demonstrates the importance of detailed consideration of practices and relations within advanced capitalist states as they actually are constituted rather than according to the dictates of abstract theories based on one or two specific countries. Any general theories of central-local or state-city relations within modern nation-states remain little more than empty generalizations based on ethnocentric assumptions until they are able to deal adequately with the exceptional cases as exemplified in Australia.

REFERENCES

Ball, M. (1983). *Housing policy and economic power.* London: Methuen.

Beed, T., Stimson, R. J., Paris, C., & Hugo, G. (1989). *Housing tenure, costs and policies in Australia.* Canberra: Canberra College of Advanced Education.

Denoon, D. (1983). *Settler capitalism.* Oxford: Oxford University Press.

Flood, J., & Yates, J. (1987). *Housing subsidies study.* Canberra: Australian Housing Research Council.

Gurr, T. R., & King, D. S. (1987). *The state and the city.* London: Macmillan.

Halligan, J., & Paris, C. (1984). The politics of local government. In J. Halligan & C. Paris (Eds.), *Australian urban politics* (pp. 1-14). Melbourne: Longman Cheshire.

Harnisch, W., Harmer, J., & Moore, C. (1986). *Innovation in Federal housing assistance: A case study of the Local Government Community Housing Program.* Paper presented at the New Zealand & Australian Regional Association Conference, University of New South Wales, November 30-December 3, 1986.

Hogwood, B. W., & Gunn, L. A. (1984). *Policy analysis for the real world.* Oxford: Oxford University Press.

Kemeny, J. (1981). *The myth of home ownership.* London: Routledge & Kegan Paul.

Kemeny, J. (1983). *The great Australian nightmare.* Melbourne: Georgian House.

Kendig, H., & Paris, C. (1987). *Towards fair shares in Australian housing.* Canberra: IYSH National Committee of Non-government Organisations.

Logan, M., Whitelaw, J., & Mackay, J. (1981). *Urbanization: The Australian experience.* Melbourne: Shillington House.

Paris, C. (1983). The myth of urban politics. *Environment and Planning D: Society and Space, 1*(1), 89-108.

Parkin, A. (1982). *Governing the cities.* Sydney: Allen & Unwin.

Power, J., & Halligan, J. (1987, September/October). Changing patterns of government in Australian metropolitan areas. *Local Government Studies, 13,* 77-86.

4

National Housing Policy and
Local Politics in Sweden

LARS NORD

SWEDEN'S HOUSING POLICY is a manifestation of comprehensive welfare ambitions (recent studies in English include those by Anas, Jirlow, Gustafsson, Hårsman, & Snickars, 1985; Heclo & Madsen, 1987; Lundqvist, 1987, 1988). According to the Swedish Constitution, the community is obliged to guarantee the rights to work, to education, and to housing at the same time as it must promote social care and security and a favorable living environment (Swedish Riksdag, 1981).

Responsibility for welfare policies is shared by central and local government (intermediate levels of government play a crucial role only in health care; Magnusson & Lane, 1987, p. 14). The division of work between the two main tiers of government, however, varies from service to service. In housing, both levels play important roles. The lion's share of the economic expense rests upon the national government in the form of housing allowances and subsidized loans for production. Local government is responsible for planning and implementation of building. Housing allowances for the elderly constitute a salient exception to this general pattern. Here, municipalities have a decisive influence upon the size of the subsidy, although national government gives some financial support.

Is this role of local government compatible with the welfare commitment of the Swedish Constitution? In this chapter, I will examine this question through an empirical analysis of 35 municipalities with at least 50,000 inhabitants in 1980, together almost one-half of Sweden's population. The question also has relevance for the more general problem of central and local government relations (e.g., Goldsmith, 1986;

Greenwood & Stewart, 1986; Hoffman-Martinot, 1987; Magnusson, 1986; Page & Goldsmith, 1987; Rhodes, 1986; Smith, 1985).

LOCAL HOUSING ALLOWANCES FOR THE ELDERLY

To help ensure a decent living standard for the elderly, subsidization of their housing costs has been a traditional part of Swedish welfare policy. Although this support has been differently shaped, the current policy involves a system of housing allowances (local housing allowances for the elderly, or LHAE) in which central and local government play different roles (see Table 4.1).

The *target group* is set by the national legislature, while the *amount of subsidy* is of local concern. *Financing* responsibility was divided in 1980. The local government began then to receive central government grants if certain conditions were fulfilled. At first, these subsidies covered 43% of the local outlays, but in 1981 they were reduced to 38% and in 1982 to 25%, where they have been ever since. The *criteria of allocation* are defined at the central level. From the outset, the Swedish Parliament decided that LHAE should be, in Titmuss's (1976, pp. 113-123) terms, selective as opposed to universal, that is, supplied only to elderly persons with "need" for support and not to the group as a whole. Income and costs of housing determine needs and the rules of calculation are given by a central authority, the National Social Insurance Board. As to *agenda-setting*, until 1988 municipalities decided by themselves whether to introduce the allowance or not. Then it became compulsory. Finally, concerning *administration*, the payment of LHAE to the individual is organized in a functional way, since the National Insurance Board and its field offices are responsible. The applicant turns to the local office in his or her municipality for LHAE, and the amount is paid monthly at the same time as the national basic pension and the national supplementary pensions scheme.

LHAE AS WELFARE SUPPORT

The significance of LHAE for the individual depends upon its position as a social right. Historically, democratic citizenship has evolved to include not only civil and political but also social elements, implying the right to economic welfare and a life according to prevailing standards (Marshall, 1975, pp. 65-122). There is an important difference, however, between civil and political rights, on the one hand, and social rights on the other. Civil and political rights are legal by nature. Their

TABLE 4.1

Responsibility of Central and Local Government Regarding LHAE

Level of Government	Target Group	Amount of Subsidy	Financing	Criteria of Allocation	Agenda-Setting	Administration
Central	X		$(X)^a$	X		X
Local		X	X		X^b	

a. Central government grants were introduced in 1980. The parentheses indicate that there are certain conditions attached to them.
b. From 1988 every municipality has been obliged to supply LHAE.

implementation is controlled by the courts and public authorities, which in principle are independent of government and legislature. In contrast, social rights are not legally binding. Their implementation is decided by political bodies. They are obligations in the form of goals, the fulfillment of which is contingent upon political will and resources available. Thus, more directly than rules of law, they are subject to changes in the political climate. They are respected as long as "the public, the politicians and the administration fully accept the legitimacy of the claims, take them seriously and give them a high degree of priority" (Marshall, 1981, p. 89).

Is this requirement met as far as LHAE is concerned? As noted, social rights are recognized in the Swedish Constitution. Any citizen can refer to this formulation to underline demand for support. However, by transferring the right to agenda-setting to the municipalities, the national government declined to take any responsibility in that respect. It was a matter of local government. Did this transfer affect the introduction of the subsidy?

In the main, the answer is no. Only one year after the parliamentary decision in 1954, all but ten municipalities had introduced the LHAE benefit (SOU, 1956, p. 32). In the early 1960s even those exceptions disappeared (SOU, 1975a, p. 328). Therefore, it seems reasonable to conclude that LHAE was quickly accepted as a social right.

However, one must also take into account the effects of the allocation criteria, set by the national government, and the amount of subsidy, set by the municipalities. These provisions identify elderly who are entitled to support. In 1987 they included about 30% of all old-age pensioners (RFV, 1987, p. 33), and the total amount of money paid out was 4,350 million Swedish crowns (i.e., about $650 million U.S.; Regeringens proposition 1986/87:100, bilaga 7, p. 53). As these figures indicate, LHAE is a significant form of welfare support.

The data so far give only an overall view. It is necessary to examine the consequences of the local right to determine the amount of subsidy by comparing different municipalities. Did such local discretion result in divergences? To begin with, the highest and lowest maximum sums paid out between 1974 and 1987 are presented in Figure 4.1, which shows considerable local variations in the maximum subsidies. The gap between the highest and lowest amount has also markedly increased since 1981. The difference in 1974 was 7,000 crowns; it was 20,000 crowns in 1987. Depending upon which municipality one lives in, LHAE's share of one's national basic pension can thus vary between 25% and 60%. LHAEs affect the disposable income of the elderly more than do variations in local taxes and in fees for domestic help (Jonsson & Tingvall, 1987, pp. 35-40).

How are such large inequalities compatible with the welfare notions laid down in the Swedish Constitution? By transferring decisions on introduction and the amount of subsidy to the municipalities, the national government declined to a great extent to adhere to public obligations. But does this mean that local government instead assumed these duties?

They would have done so if they had given allowances "to the greatest possible extent," as a royal commission dealing with constitutional social rights once postulated (SOU, 1975b, p. 184). However, this was obviously not the case, since some of them offered subsidies that were half those of others. As a matter of fact, resources play a minor role in determining the attitudes of the municipalities concerning LHAE (see below). Therefore, reliance on local government to implement social rights is an uncertain policy. Next, by analyzing some reasons for the big differences among municipalities regarding LHAE, I will discuss why this is so.

LHAE AND POLITICS

In most political systems, decision makers are influenced by socioeconomic, political, and administrative factors. These might be seen as criteria inciting public authorities to act (Hansen, 1981, p. 45). The following discussion examines these three factors.

LHAE was initiated to reduce housing costs of the elderly. The size of these costs and the share of pensioners in the population might, therefore, be criteria expressing the need for subsidies. A royal commission indeed established positive correlations among rent level, age structure of the population, and LHAE subsidies (SOU, 1977, p. 271).

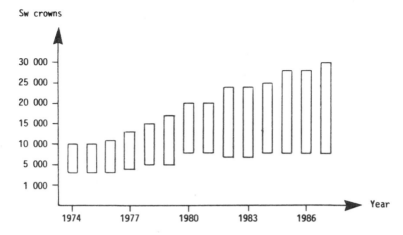

Figure 4.1. Highest and Lowest Annual Maximum Subsidies to a Single Retired Person, 1974-87

SOURCE: Official documents from the municipalities.
NOTE: In a few cases when the maximum subsidy was the same as the real rent paid, the annual mean rent for two rooms and a kitchen, 70 m^2, in privately and newly built dwelling houses with state loans, served as the basis for calculation (SCB, 1987a, Table 9.2.4; 1987b).

However, average annual rents for dwellings with two rooms and a kitchen in semipublic multistory buildings in five different kinds of municipalities in 1975, 1980, and 1985 differed so little that they cannot explain the local variations in LHAE subsidies (SCB, 1976, p. 10; 1981, p. 13; 1986). These results are also supported by my own data from 1985.

The age structure of the population might have two types of effects. The fewer old people, the more one can afford to subsidize them. But the more old people, the more important they are as a pressure group for demanding LHAE. Earlier Swedish studies have shown that, in general, public funds for care of the elderly do not increase with age of the population (Johansson, 1982, p. 35; Stjernquist & Magnusson, 1988, p. 173). This pattern is confirmed by my results. Municipalities with a young population do not offer higher LHAE than those with more old people.

Conceivably, availability of resources could explain variations in LHAE. Since, in 1983, for example, 9% of municipal outlays for the elderly were LHAE costs, economic considerations should be of a certain significance in determining the maximum of LHAE given.

Local taxes are the most important source of revenue for the municipalities, in 1985 amounting to 41% of total income (SCB, 1988, p. 260). Therefore, it seems relevant to investigate the correlation between LHAE and the tax capacity of the population, the latter being the tax basis divided by the number of inhabitants.

Earlier studies show contradictory results (Johansson, 1982, p. 30; Stjernquist & Magnusson 1988, p. 172). The present study found a decreasing positive correlation between tax capacity and LHAE levels for the three years studied. In 1975, 28% of the variance was explained by differences in tax capacity. In 1980, this was down to 23%, and in 1985, it was only 16%.

Since less and less of LHAE divergences are due to differing tax capacities, I next tested the importance of another source of revenue: central government grants, introduced in 1980 to equalize LHAE variations between the municipalities (Regeringens proposition 1978/ 79:95, p. 76). As shown in Figure 4.1, these intentions were attained only for a short time. In 1982, the variations were the same as before. As their share was reduced from 43% to 25% of the municipal expenses during these years, these grants were probably too modest to achieve their purpose.

We now turn to political/ideological factors. Their importance is unclear. In Denmark, for example, the Social Democrats have had a marked influence on the level of local expenditure concerning homecare payments for the aged (Skoovsgaard, 1981, pp. 55ff.). In Great Britain, as well, Labour advocates better services for the elderly (Davies, Barton, McMillan, & Williamson, 1971, pp. 133ff.). In Sweden, studies point in diverse directions. It has been stated that socialist majorities are more inclined than bourgeois ones (Conservative, Liberal, and Farmers' parties) to spend money on social issues (Johansson, 1982, pp. 42-44; Olander, 1984, pp. 243ff.), but also that the color of the parties in power makes no difference (Murray, 1981, p. 271). The same contradictory conclusions are found regarding LHAE (Johansson, 1982, p. 30; Murray, 1981, p. 238).

My own study examined political parties as exponents of divergent political ideologies. Results show no relationship between the color of the ruling party bloc (Social Democratic and Communist parties on the one hand and the Conservative, Liberal, and Farmers' parties on the other) and maximum LHAE levels in 1975, 1981, and 1987.

Another way of ascertaining the importance of political factors is to study the effect of interparty competition. Earlier research has shown that the more parties compete with each other, the more similar the

TABLE 4.2

Political Majorities and the Growth of LHAE, in Percentages, 1975-1987

Political Majority	Percentage Increase of LHAE Maximum	N
Change		
socialist to bourgeois	24.6	10
bourgeois to socialist	28.2	14
No change		
bourgeois majority	32.5	21
socialist majority	32.1	59
unclear majority	39.1	29

expenditure patterns, irrespective of party control (Sharpe & Newton, 1984, p. 200). An earlier study of the municipal elections of 1968 and 1976 could not attribute increases in the LHAE subsidy to competition (Johansson, 1982, p. 30). My own results, based on four elections between 1976 and 1985, are presented in Table 4.2, which relates the raising of the maximum level of LHAE after each election to the distribution of seats before and after the election.

It seems that competition limits increases of LHAE when there is a change of power from one political bloc to another and promotes increases when the political situation is unclear, and with stable majorities. Thus the effects of competition depend on its form. One might explain the difference by the relative importance of LHAE as a political issue. When there was such a huge swing in the electorate that the majority changed, LHAE played a minor role. When neither side succeeded in tipping the political balance in its favor, every single issue was considered important. The elderly might then have a decisive influence on the outcome of the election, so that it became necessary to pay attention to the LHAE level.

In sum, socioeconomic factors such as housing costs, share of elderly persons in the population, and economic resources clarify local variations in LHAE as little as political factors such as party ideology and interparty competition. Next, we examine administration of LHAE.

LHAE AND ADMINISTRATION

The administration of LHAE is handled by a central authority and its field offices. The municipalities have no formal power whatsoever to interfere with the implementation of decisions about standards. This study confines itself to the political process *preceding* the decisions made by political bodies. Municipal officials responsible for preparing LHAE decisions were interviewed to investigate factors that might have influenced their work in this respect.

Among the many administrative factors that may affect the administrator, institutional circumstances are the focus here. The very complicated construction of LHAE, which leaves much room for expertise and technical competence, is enough to justify this approach. In addition, since the implementation of local decisions as to LHAE is in the hands of officials employed by the national government, local politicians have no formal right to supervise an important link in the chain of decisions characterizing a political process. Many of the channels for control and communication, as well as for transferring authority, necessary for internalizing the goals of an organization among the personnel (Simon, Smithburg, & Thompson, 1964) do not thus exist. Those material and symbolic means that, according to Etzioni (1975, p. 5), are employed to make subjects comply to directives and other norms are missing. The local politicians have no right to intervene with respect to, for example, the allocation of salaries and wages, of esteem and prestige symbols, or of "fringe benefits" for those working to execute what they have decided. The role of those professionals preparing the political decisions is correspondingly strengthened.

The first institutional factor to be investigated is incrementalism, an approach whereby decision makers take only factors of immediate concern to themselves into consideration, leading to slow, small changes in the status quo instead of drastic alterations. Available means, not clear-cut, fixed objectives, determine outcomes. A policy is "tried, altered, tried in its altered form, altered again, and so forth" (Braybrooke & Lindblom, 1963, p. 73). "The science of muddling through" is a phrase catching the partial, partisan, and multiplex character of the political process according to this thinking (Lindblom, 1959, 1979).

In this study, incrementalism was measured by comparing the 35 municipalities in 1975, 1981, and 1987 with regard to how much they had changed their LHAE maximum levels in terms of intervals of 3,000 Swedish crowns (10% of the highest maximum sum in 1987). There was no or one interval change in 16 cases, changes between two intervals in 8 cases, and changes between three intervals in 11 cases. In other words, small increases have been preferred to large ones, and the initial subsidy level was determinative, supporting the notion of incrementalism. It is not clear, however, how these initial levels were established.

TABLE 4.3

Average LHAE Maximum Levels in Local Departments for
Social and Nonsocial Services for Five Different years, 1974-1987

Department of Administration	LHAE Maximum Sums					N
	1975	1978	1981	1984	1987	
Social Services	6,590	9,780	13,370	15,010	18,230	23
Other	5,500	7,930	11,410	14,020	15,370	12

Another administrative factor that might influence LHAE decisions is the kind of municipal branch responsible for LHAE administration. In 23 municipalities, LHAE was handled by social administrators, and in 12 by officials from departments for finance, central administration, or real estate.

Are social administrators more inclined to look after the interests of the elderly than officials dealing with matters of economy and general administration? Table 4.3 shows the average maximum levels in Swedish crowns for both categories for five different years. The contrast is striking. In all years, municipalities where LHAE has been handled by the department especially devoted to social issues have been more generous than those where the allowance has been in the hands of professionals not working with welfare provisions.

Regarding the growth of LHAE shown in Table 4.4, the figures are more equivocal: Social officials no longer come forth as the uncontested champions of the interests of the elderly. As a matter of fact, between 1976 and 1984, they were thoroughly pushed into the background by their colleagues working with nonsocial issues. Why? How can we explain this divergence?

An answer may lie in the fiscal crisis in municipalities that restrained the propensity for expansion and expenses at the end of the 1970s and the beginning of the 1980s in Sweden, as in many other countries. In situations of retrenchment, officials working with social services are in a weaker position than those who are responsible for economy and capital assets. The former are supplicants before the latter, who control and have an overall view of the finances. To conclude, the findings imply that social officers are more likely than nonsocial officers to fight for welfare provisions. However, their capacity is reduced in economically hard times.

TABLE 4.4

Average Growth in percentage of LHAE Maximum Levels in Departments
for Social and Nonsocial Services During Periods of Three years, 1975-1987

Department of Administration	Growth in Percentage of LHAE Maximum				
	1975-78	1978-81	1981-84	1984-87	N
Social services	53	40	11	25	23
Other	50	49	23	11	12

It may be added that attempts to find additional correlations in the
material available have proved fruitless. Thus incrementalism is related
neither to any kind of municipality nor to any type of administrator.
That departments of social services were in the majority in the group of
municipalities with most interval changes, 8 of 11, is not much proof to
the contrary.

In sum, this study of the consequences of transferring responsibility
for housing policy to local government suggests that administrative
factors (incrementalism, municipal branch responsible for preparing
LHAE decisions) are more important than socioeconomic and political
factors for variations among municipalities with regard to housing
allowances for the elderly.

CONCLUSION:
LHAE AND LOCAL GOVERNMENT

To distribute a subsidy according to universal principles is the most
egalitarian way to carry out welfare obligations. Then all people within
a certain category benefit equally in accordance with their collective
needs. If the subsidy is selectively supplied, based on the needs of the
individuals, there is a departure from universality, but all people are
affected equally, irrespective of their place of residence. This is not the
case when the national government has transferred the right of allocation to local government. Then, the political/administrative jurisdiction
may affect distribution. To what extent the consequences are detrimental to citizens' ability to receive their social rights is an empirical
question.

This chapter has presented results of such an empirical investigation.
The distribution of LHAE among the elderly, owing to municipalities'
right to determine their maximum subsidy levels, was very uneven in

the period 1974-1987. Depending upon where an elderly individual lived, the amount of subsidy given influenced to a considerable degree his or her disposable income. This is a clear violation of Sweden's constitutional commitment to social rights. Considering the large variations in LHAE levels, the subsidy was not provided "to the greatest possible extent," as a royal commission once found reasonable (see above).

The decisions of the municipalities with regard to LHAE are, as this study shows, decisively influenced by neither socioeconomic nor political/ideological considerations. Instead, it seems that administrative factors are the most important. Thus, if the central government grants were increased, it is not sure that an equalization of the levels would occur. Certain municipalities with the lowest subsidies and the lowest tax capacities, would, perhaps, increase the allowance if the national government offered more grants, but others probably would not. National government has now made LHAE compulsory, but it still permits municipalities to determine the maximum sum above a certain level, which is so low that most of the municipalities investigated in this study had already surpassed it when the obligation was introduced in 1988. Raising the minimum level would, on the other hand, make a mockery of local government. However, it appears that this has to be done if the welfare obligation is to be fulfilled. The logical outcome is to revert the right to determine the amount of LHAE subsidy to national government.

REFERENCES

Anas, A., Jirlow, U., Gustafsson, J., Hårsman, B., & Snickars, F. (1985). The Swedish housing market: Structure, policy and issues. *Scandinavian Housing and Planning Research, 2*, 169-187.

Braybrooke, D., & Lindblom, C. E. (1963). *A strategy of decision.* New York: Free Press.

Davies, B. P., Barton, A. J., McMillan, I. S., & Williamson, V. K. (1971). *Variations in services for the aged.* Birkenhead: Willmer Brothers.

Etzioni, A. (1975). *Comparative analysis of complex organizations.* New York: Free Press.

Goldsmith, M. (Ed.). (1986). *New research in central-local relations.* Aldershot: Gower.

Greenwood, R., & Stewart, J. P. (1986). The institutional and organizational capabilities of local government. *Public Administration, 64*, 35-50.

Hansen, T. (1981). Transforming needs into expenditure decisions. In K. Newton (Ed.), *Urban political economy* (pp. 27-47). London: Frances Pinter.

Heclo, H., & Madsen, H. (1987). *Policy and politics in Sweden: Principled pragmatism.* Philadelphia: Temple University Press.

Hoffman-Martinot, V. (1987). *Finances et pouvoir local: l'expérience allemande.* Paris: Presses Universitaires de France.

Johansson, L. (1982). *Kommunal servicevariation* [Municipal variations in services] (Ds Kn 1982:2). Stockholm: Liber.

Jonsson, A., & Tingvall, L. (1987). *Det handlar inte bara om kommunal skatt* [It is not only about local taxes]. Stockholm: Svenska Kommunförbundet.

Lindblom, C. E. (1959). The science of muddling through. *Public Administration Review, 19*, 79-88.

Lindblom, C. E. (1979). Still muddling through. *Public Administration Review, 39*, 517-526.

Lundqvist, L. J. (1987). Sweden's housing policy and the quest for tenure neutrality. *Scandinavian Housing and Planning Research, 4*, 119-133.

Lundqvist, L. J. (1988). *Housing policy and tenures in Sweden*. Aldershot: Gower.

Magnusson, W. (1986). Bourgeois theories of local government. *Political Studies, 34*, 1-17.

Magnusson, T., & Lane, J.-E. (1987). Sweden. In E. C. Page & M. J. Goldsmith (Eds.), *Central and local government relations: A comparative analysis of West European unitary states* (pp. 12-28). London: Sage.

Marshall, T. H. (1975). *Class, citizenship, and social development*. Westport, CT: Greenwood.

Marshall, T. H. (1981). *The right to welfare*. London: Heinemann.

Murray, R. (1981). *Kommunernas roll i den offentliga sektorn* [The role of the municipalities in the public sector]. Stockholm: Nationalekonomiska institutionen, Stockholms universitet.

Olander, L.-O. (1984). *Staten, kommunerna och servicen* [The state, the municipalities, and the services] (Ds C 1984:5). Stockholm: Liber.

Page, E. C., & Goldsmith, M. J. (Eds.). (1987). *Central and local government relations: A comparative analysis of West European unitary states*. London: Sage.

Regeringens proposition (government bill). (1978/79). 95.

Regeringens proposition (government bill). (1986/87). 100, bilaga 7.

RFV (National Social Insurance Board). (1987). *Rapport. RFV 1987:6*. Stockholm: Author.

Rhodes, R. A. W. (1986). *The national world of local government*. London: Allen & Unwin.

SCB (Central Bureau of Statistics). (1976). *Statistiska meddelanden. Bo 1976:10* [Statistical bulletins]. Stockholm: Author.

SCB (Central Bureau of Statistics). (1981). *Statistiska meddelanden. Bo 1981:13* [Statistical bulletins]. Stockholm: Author.

SCB (Central Bureau of Statistics). (1986). *Statistiska meddelanden. Bo 30 SM 8601* [Statistical bulletins]. Stockholm: Author.

SCB (Central Bureau of Statistics). (1987a). *Bostads- och byggnadsstatistisk årsbok 1986/87*. Stockholm: Author.

SCB (Central Bureau of Statistics). (1987b). *Statistiska meddelanden. Bo 30 SM 8701* [Statistical bulletins]. Stockholm: Author.

SCB (Central Bureau of Statistics). (1988). *Statistisk årsbok 1987*. Stockholm: Author.

Sharpe, L. J., & Newton, K. (1984). *Does politics matter?* Oxford: Clarendon.

Simon, H. A., Smithburg, D. W., & Thompson, V. A. (1964). *Public administration*. New York: Knopf.

Skoovsgaard, C.-J. (1981). Party influence on local spending in Denmark. In K. Newton (Ed.), *Urban political economy* (pp. 48-62). London: Frances Pinter.

Smith, B. C. (1985). *Decentralization*. London: Allen & Unwin.

SOU (Swedish Government Official Reports). (1956). *SOU 1956:1*.

SOU (Swedish Government Official Reports). (1975a). *SOU 1975:51*.

SOU (Swedish Government Official Reports). (1975b). *SOU 1975:75.*
SOU (Swedish Government Official Reports). (1977). *SOU 1977:78.*
Stjernquist, N., & Magnusson, H. (1988). *Den kommunala självstyrelsen, jämlikheten och variationerna mellan kommunerna* [Local government, equality and variations among the municipalities]. Stockholm: Civildepartementet.
Swedish Riksdag (1981). *The Constitutional Documents of Sweden: The instrument of government.* Stockholm: Author.
Titmuss, R. M. (1976). *Commitment to welfare.* London: Allen & Unwin.

Part II

Rental Housing

Introduction

CHESTER HARTMAN

ALTHOUGH THE PROPORTIONS of rental units in the United States and Britain are roughly the same, the rental sectors in the two countries could hardly be more dissimilar. In the United States, well over 90% of the rental stock is privately owned and operated. In Britain, nearly three-fourths of the rental stock is owned and operated by local public authorities, and an additional one-seventh is controlled by private nonprofit housing associations, a type of entity that plays an extremely minor role in the United States. (While quite a number of nonprofit, often community-based, housing development corporations exist in the United States, few produce or manage more than a trivial number of units; by comparison, seven housing associations in England manage 10,000 units or more—a figure no similar entity in the United States comes close to matching.) The roles of the private and public sectors and the prevailing beliefs underlying these roles are completely opposite in the two societies—the inroads of Thatcherism notwithstanding.

Furthermore, Britain has a housing policy (even if, under the Thatcher government, it is one aimed at diminishing and weakening the public sector), whereas almost no one would argue that the United States has anything that resembles a comprehensive, coherent housing *policy* (unless by that term is meant the eclectic assemblage of government programs, minuscule in the aggregate, that comes and goes in seeming random fashion, both from administration to administration and within the tenure of a given administration, plus the historically favorable treatment home ownership gets under the U.S. income tax system).

And because Britain has a housing policy, as well as some sense of a national development policy, also virtually unknown in the United States, the issue of labor mobility as it relates to housing availability can be addressed there: attempting to facilitate movement of workers from areas of high unemployment to areas of labor shortage by making affordable housing available. (By way of contrast, in the Washington, D.C., metropolitan area, for example, sorely needed construction workers travel up from southern Virginia and spend weekday nights camped in recreational areas; an "affordable housing task force" in suburban Fairfax County, Virginia, issued as its principal recommendation a proposal to extend the permitted stay of such workers in park camping grounds beyond the short-term duration set with recreation users in mind.)

Because rental housing is and has been so major a part of the government's enterprise in Britain, housing issues there take on a public and political importance unknown in the United States. Council estate (public housing) residents are, as Kleinman notes in Chapter 5, "the heartland of Labour party support," and, as such, "provide an impediment to the spread of the Thatcher revolution to all parts of the country." Thus the move to weaken this sector and replace it with a strong private rental sector is an element of a broader political strategy. In the United States, by contrast, not only is the public housing population tiny (about 1.5% of the total population), but, because of its income, race, and age composition—half of all residents are under the voting age of 18—it represents an even smaller percentage of registered voters and those who actually vote. The markedly different ways in which the two governments treat the homeless and the problem of homelessness also are a function of the broader role of the government-supported rental housing sector and of local government housing agencies in Britain. In the United States, local housing authorities play little role in housing the homeless, and the problem is regarded by and large as one of temporary or transitional shelter and supportive services for problem families and disabled individuals, rather than as fundamentally a housing problem. In England and Wales, local authorities accepted responsibility for 118,000 homeless families in 1987 alone.

More generally, national policies in Britain hold sway, in housing as well as other areas, in contrast to the high degree of decentralization and localism that characterizes the U.S. housing system and the political system as a whole. While in the United States the national administration can exercise some political pressure on localities to alter regulatory controls, for the most part such decisions are made in a fragmented way at the local level. In both systems, however, national subsidization policies—either directly or indirectly via the tax system—deeply influence how much rental housing gets built, by whom, and for whom.

In both countries, it is clear that rental housing affordable by lower-income consumers will not be built without some form of substantial direct or indirect government assistance.

In both the Thatcher and Reagan-Bush administrations, efforts are being made to sell off the public housing stock to tenants—a move motivated largely by ideological considerations and attracting support from the tenants, who are able to purchase their units at bargain "insider" prices. In Britain, some 1 million units—about one-sixth of the total stock—have already been sold off in this way, with plans to sell off an additional 100,000 units a year in the 1990s. In the United States, the pace is far slower, in large part because relatively few units lend themselves to individual sale, in terms of design and condition, and in large part because the incomes of public housing tenants generally are too low for them to afford even subsidized home ownership. (There are a very few successful examples of such sales in the United States, most notably in St. Louis.) The Reagan and Bush administrations and the current HUD secretary, Jack Kemp, are giving considerable publicity to a Washington, D.C., project, Kenilworth-Parkside, as a shining example of tenant ownership (not yet achieved there) and tenant management (a quite different, far more promising notion), although a recent General Accounting Office study reported that the costs of the modernization program under way at this "show" project in order to prepare it for eventual sale will be approximately $70,000 per unit, seven times the national average for public housing modernization and over twice the per unit cost of other D.C. projects undergoing modernization.

Beyond programs to sell off public housing to tenants, the other major potential loss to the stock of government-assisted low-rent housing in the United States comes from the so-called prepayment problem that Appelbaum and Dreier describe in Chapter 6: In the absence of either federal controls or additional subsidies, tens of thousands of subsidized rental units may be lost, as their private owners disengage from these subsidy programs and concomitant controls in order to take advantage of profitable market conditions.

Public housing construction in the United States has virtually come to an end; in Britain, even with sharp cutbacks in housing expenditures by the Thatcher government and a shift in expenditures from new construction to rehabilitation, some 13,000 new public housing units nonetheless were started in 1988. Thus some of the demand for public housing in Britain from families on the waiting lists can be met from new construction, whereas in the United States virtually all such demand is met from relets. Waiting lists for public housing in most U.S.

cities are huge, with waiting periods of 5-10 years and longer the norm. There is little mobility out of public housing in the United States, except among the elderly, by death or removal to a nursing facility. Families tend to remain in these quarters permanently, with two- and even three-generation public housing families not at all uncommon due to permanently low incomes and the small and shrinking supply of decent, affordable accommodations in the private market. While Kleinman does not present data on the size of waiting lists or length of waiting period, the fact that over 200,000 units become available annually via the reletting process indicates a far greater openness in the British system, with respect to both exit and entry.

Quality problems and modernization needs are rife in the public housing stocks of both nations. Kleinman cites a 1985 government study showing an estimated backlog of £20 billion; a more recent study done for HUD shows a relatively far larger per unit $22 billion backlog for the U.S. public housing program.

Whether the privatization agenda of the Thatcher and Reagan-Bush governments continues over the immediate future is a function in part of whether, in Great Britain, where governments have nonfixed terms, the Thatcher revolution is permitted to continue, and in part of objective housing conditions, which in both countries are deteriorating, as measured by affordability problems and the outrage of homelessness. Political demands for a more responsive, effective government role with respect to that one-third of the population in both countries that rents rather than owns will determine the future of rental housing in both societies.

5

The Future Provision of
Social Housing in Britain

MARK KLEINMAN

HOUSING POLICY IN BRITAIN toward the rental sector is currently undergoing a major change. The Conservative government's white paper, *Housing: The Government's Proposals*, published in September 1987, and the subsequent 1988 Housing Act and 1989 Local Government and Housing Act mark a decisive break with the consensus policies that have been pursued through most of the postwar period. Since 1945, governments of both parties have subscribed to the view that council housing—the direct provision of rental housing by local authorities—was the major method by which the need for rental housing would be met. However, the present government has broken with this consensus. In the future, local authorities are no longer to be providers of housing, but, in the words of the white paper, "should increasingly see themselves as enablers who ensure that everyone in their area is adequately housed—but not necessarily by them" (Department of the Environment/Welsh Office 1987).

The Conservative government elected in 1979 has never been a supporter of council housing. In its first two terms, the Thatcher government's antipathy toward council housing was expressed mainly in terms of a strongly pro-owner occupation policy. In particular, council tenants were given the right to buy their existing council homes. Since

AUTHOR'S NOTE: This chapter is a revised version of a paper presented at the International Conference on Housing, Policy and Urban Innovation, Amsterdam, June 27-July 1, 1988. It is based on research carried out at the University of Cambridge, Department of Land Economy, and supported by the Joseph Rowntree Memorial Trust as part of its Housing Finance Programme.

1980, more than a million dwellings have been sold in this way (Forrest & Murie, 1988).

In the third term of the Thatcher government, the emphasis of housing policy has shifted (Monk & Kleinman, 1989). Although the first objective of the government remains to spread home ownership as widely as possible, attention has now been focused on the problems of the rental sector. This has happened for four reasons.

First, there has been a growing realization on the part of government that owner occupancy alone cannot provide a complete housing solution, and a considerable proportion of the population at any one time will be tenants.

Second, there has been increasing interest in the relationship between housing markets and labor mobility (Bover, Muellbauer, & Murphy, 1988; Hughes & McCormick, 1987; Minford, Peel, & Ashton, 1987). Economic recovery in Britain has largely been concentrated in the Southeast. Shortages of labor in that region coexist with high unemployment elsewhere, and, in the view of government, rigidities in the housing market are partly to blame. This has been accentuated by the gap in house prices between North and South. In mid-1988, data from the Halifax Building Society showed that a three-bedroom semi-detached house cost almost *four times* as much in Greater London as in South Yorkshire. Greater supply of rental units in the growth regions will, in the government's view, enable unemployed workers to move to take up jobs.

Third, housing problems have, to some extent, become more of a public issue in recent years, particularly the rise in homelessness. The numbers of homeless households rehoused by local councils has doubled in ten years; in 1987, local authorities in England and Wales accepted responsibility for 118,000 homeless households (Central Statistical Office, 1989, p. 139).

Finally, there are political factors. Council estates are seen as being the heartland of Labour party support and, as such, as providing an impediment to the spread of the Thatcher revolution to all parts of the country. The diversification of the rental housing stock among a range of types of landlord is seen as generating political benefits in weakening the umbilical connection between municipal provision and Labour voting (Kemp, 1989; Selbourne, 1987).

The 1987 white paper and the 1988 Housing Act therefore place a great deal of emphasis on the importance of putting "new life into the independent rental sector." This strategy has two aspects, involving both profit-seeking private landlords and nonprofit housing associations. First, by deregulating new lettings, allowing landlords to charge market rents, and weakening tenants' security of tenure, the government hopes to reverse the long-term decline of the private rental sector and to encourage new investment by private landlords. Investor response to deregulation is discussed in more detail by Crook, in Chapter 11 of this volume.

Second, the government is seeking to expand the role of housing associations (Hills, 1987; Kemp, 1989; Kleinman, 1987).[1] Hitherto, the associations have been publicly financed and have received both generous capital subsidies and, where necessary, a "safety net" of revenue support. In the future, associations will receive lower amounts of public subsidy than at present, and will be encouraged to use private in addition to public finance.

This will allow associations to expand somewhat, but they will have to bear some of the commercial risk of undertaking development. Rents will be below market levels (because of public subsidy), but will be higher than those that associations currently charge. In addition, the 1988 Housing Act contains mechanisms encouraging the transfer of existing tenanted council properties to alternative landlords, the vast majority of whom are likely to be housing associations.

In this chapter, I look at what has happened to the supply of rental housing becoming available for new tenants seeking to enter the local authority sector over the last ten years or so. Despite the current policy interest in the independent rental sector, I concentrate on the local authority sector for two reasons. First, local authorities remain by far the major suppliers of rental housing in Britain (see Table 5.1); they own almost three-quarters of the rental stock in Britain and, even in a period of public expenditure cutbacks, they contributed nearly two-thirds of new completions between 1984 and 1986. Second, they are likely to remain the major source of rental accommodation to meet housing needs until the end of the century, even on the most optimistic assessment of the current government's attempts to revive the independent rental sector.

THE SUPPLY OF RENTAL HOUSING
AND ITS POLICY DETERMINANTS

The analysis of the supply of rental housing becoming available for those seeking to obtain public rental housing is a neglected area of research. Considerable attention has been given, on the one hand, to the issue of public sector output and how this is influenced by central

TABLE 5.1

The Rental Sector in Britain

	Dwellings, 1988		
	Number *(in thousands)*	*% of Rental Stock*	*Average Completions* *1986-1988*
Local authorities[a]	5,606	72	20,400
Housing associations	566	7	11,700
Private landlords	1,606	21	—
Total	7,778	100	32,100

SOURCE: Department of Environment, housing and construction statistics.
a. Includes New Town Corporations.

government policies and, on the other, to questions relating to the selection of tenants and the allocation of dwellings. But the crucial links between these two processes have been relatively neglected in the literature.

Supply for new tenants consists of two elements—new building and relets of the existing stock. New construction always creates an extra unit of supply. Either the new unit is let directly to a starter to the sector or it is let to an existing tenant, thereby creating (perhaps through a chain of such moves) an extra unit for letting to a starter. However, not all relets create new vacancies—relets created by transferring or exchanging tenants do not. Only where the relet is caused by the departure of an existing tenant from the sector, the death of a tenant, or the dissolution of a tenant household will a net vacancy arise. Thus total supply to new tenants is equal to the sum of new build plus net relets. In this chapter the term *relets* will refer to *net* relets.

Over the past decade, two policies in particular have influenced the size of the supply to new tenants. First, there has been the reduction in the size of the capital expenditure program after 1976-77 and the shift within the budget toward rehabilitation of the existing stock. As a result, new housing starts by local authorities in England declined from 108,000 in 1976 to only 13,000 in 1988. Second, the introduction of the "right to buy" led to sales of more than a million dwellings in England and Wales between 1980 and 1988. This in turn led to an actual decline in the size of the council sector, in both absolute and relative terms.[2] In 1980 there were 6.6 million council dwellings in Britain, representing 31.1% of the housing stock. By 1988 this had fallen to 5.6 million, 24.8% of the stock. A smaller stock size will mean, *ceteris paribus*, a

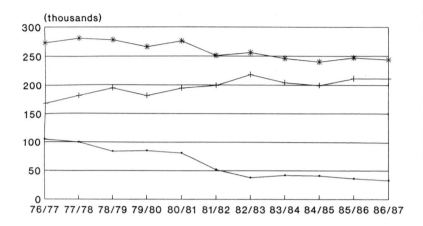

Figure 5.1. Lettings to New Tenants, England

SOURCE: Figures derived by aggregating data from the individual Housing Investment Programme returns.

smaller number of relets per annum. Thus both components of supply—new build and relets—have been reduced as a result of current housing policies.

THE SUPPLY OF RENTAL HOUSING TO NEW COUNCIL TENANTS IN ENGLAND, 1976-1986

The actual supply of rental housing to new tenants in England between 1976-77 and 1986-87 is shown in Figure 5.1. The numbers given were derived by aggregating data from the individual Housing Investment Programme returns made annually by the 366 local authorities in England (for further details, see Kleinman, 1988). Total lettings to new tenants peaked in 1977-78. By 1984-85, lettings were 15% lower than in 1977-78, although there has been a slight recovery since then.

The decline has been most severe on the new build side, with supply from this source falling by nearly 70% over the observed period. Until 1982-83, the decline in new build supply was partly offset by continuing growth in relets. Relets grew for two reasons. First, the size of the council sector as a whole continued to grow throughout the 1970s, reaching a peak in 1981 (Central Statistical Office, 1989). Second, 1980 and 1981 were difficult years for first-time buyers seeking to enter the market, mostly because of high mortgage rates (Kleinman & Whitehead, 1988). It is therefore likely that some moves from the council to the owner-occupied sector were delayed until 1982-83.

Since 1982-83, however, relets have also begun to decline (albeit somewhat erratically), reflecting the decline in the size of the council stock. From a peak of 218,000 in 1982-83, relets fell to 199,000 in 1984-85 before recovering slightly to 211,000 in 1985-86.

Thus, although policy has negatively affected both components of supply, the decline in new build has been both more important and more long-standing than the reduction in relet supply. Moreover, the sharp reduction in new build supply in the past ten years has meant that relet supply has become of increasing importance in meeting the need for rental housing. Whereas in the mid-1970s only about 60% of local authority supply came from relets, by 1986-87, more than 85% of lettings to new tenants originated in a departure from the stock.

SPATIAL ASPECTS OF THE DECLINE
IN LOCAL AUTHORITY SUPPLY

The decline in supply of local authority rental housing has not been evenly distributed across the country. Table 5.2 shows how it has changed in the four broad regions of England. It is clear that the decline has been most severe in London. Between 1980-81 and 1986-87, total supply in London fell by more than a third, from 50,000 allocations per year to just over 30,000. Even in comparison with the other major British metropolitan areas, London is shown to have been disproportionately affected by the decline in supply (Kleinman, 1988, Table 2). The reason for this lies in the relative importance in the capital of new completion in the 1970s. While in the other conurbations less than one-third of supply in 1976-77 came from new construction, in London the proportion was more than 50%.

This aspect of the decline in supply is particularly perverse. London is the area of the country where problems of access to housing are most acute. This is shown by the high levels of homelessness and the even higher numbers of households living in "nontenure" accommodation—hostels, bed-and-breakfast hotels, short-life housing, squats, sleeping on friends' floors, and so on (Conway, 1984; Kleinman & Whitehead, 1988). On grounds of social equity, then, in a period of declining supply, London should have been the area where access opportunities were most protected; in practice, the reverse occurred.

TABLE 5.2

Letting to New Tenants by Region

	Total Supply			
	1980-81	*1986-87*	*Absolute Change*	*Percentage Change*
	(in thousands)		*(in thousands)*	
North	106.9	102.5	−4.4	−4.1
Midlands	57.1	54.9	−2.2	−3.9
South	61.5	54.6	−6.9	−11.2
London	50.1	31.7	−18.4	−36.7
All areas	276	244	−32	−11.6

NOTE: North = North, Northwest, Yorkshire, and Humberside; Midlands = West and East Midlands; South = Southwest, Southeast (excluding London), and East Anglia; London = Greater London

However, from the market-oriented, laissez-faire viewpoint of the New Right, these geographical imbalances do not provide a ground for activist state intervention. Public policy should seek to accommodate rather than counter market forces. That is, if access problems are concentrated in London, while other regions or areas have crude surpluses of housing, policy should not seek to provide for such "excess demand" in London, but rather should encourage (in practice, constrain) such households to move elsewhere, particularly if they are outside the labor market and therefore do not "need" to be in London. This position has not, so far, been explicitly adopted by any British government. On the contrary, the use of supposedly objective measures of housing need, such as the generalized needs index (GNI) and housing needs index (HNI), to allocate capital resources reflected a policy of attempting to meet housing needs wherever they arose.[3]

This consensus position may now be changing. In Bristol on August 28, 1987, in a speech titled *Some Reflections on Housing Policy,* then Housing Minister William Waldegrave argued that there is a "hugely important though largely unarticulated policy of 'mixed communities' for which we are trying to pay," and cast doubt on the wisdom of paying to keep poorer people in more expensive areas. So far, this approach has not been directly translated into government housing policy. It is, however, possible that in future, government may take a more hard-line view about where rich and poor should live.

NEW SOURCES OF SUPPLY

According to the government's view, concern with the supply of council housing and with the spatial aspects of that supply is misplaced. In the future, a revived independent rental sector will take over direct provision from local authorities.

However, the measures in the 1988 Housing Act aimed at reviving an independent rental sector are unlikely to lead to any sustained increase in net new investment (see also Crook, Chapter 11, this volume). As far as private landlords are concerned, the proposals in the bill concentrate on the supply side, by seeking to increase the return on rental housing as an investment. But any revival of a true private rental sector will also have to overcome demand-side problems. Overall, demand for private rental housing is low. Most existing and potential tenants are simply too poor to be able to afford the sort of rents that would give a return on investment acceptable to landlords (see Chapter 11; also see Whitehead & Kleinman, 1986). Those tenants (or potential tenants) who could afford to pay such rents would be better off financially, and would receive more choice and greater value for their money, in the owner-occupied sector.

The only exceptions to this pattern are the relatively small groups of households who both can afford to pay market rents and, because of their particular circumstances, have some preference for the attributes of private renting (such as easy access procedures). Such households will include, for example, mobile employed households who need to change location frequently for career reasons, and perhaps some better-off elderly households who do not want the responsibilities of property ownership (Whitehead & Kleinman, 1986, 1989). Recent starters in the sector tend to be young, employed (often in nonmanual occupations) and with reasonable purchasing power, in contrast to long-term residents of the sector, who tend to be old, poor, and outside the labor force (London Research Centre, 1988; Todd, 1986).

Outside of these specific groups, there is no prospect of any increase in mainstream demand for private renting, unless fiscal subsidies were to be substantially shifted away from owning and toward renting—a policy that does not seem to be politically viable in Britain today. Hence, as argued also by Crook (Chapter 11), any increase in provision by "true" private landlords is likely to be both limited in scale and mainly targeted at groups other than those in most need.

The proposals with respect to housing associations hold out the tempting prospect for them of a larger program of activity. Nevertheless, three factors in particular suggest that the impact of the government's proposals here will tend to be more gradual than is often suggested. First, there is the question of the size of individual associations. Although there are a small number of very large associations, the majority are very small organizations. In 1983, out of the 1,577 associations registered with the Housing Corporation, there were only 7 with more than 10,000 units, while 1,140 managed fewer than 100 units each (Hills, 1987). The majority of housing associations will not be able "overnight" to expand greatly in terms of absolute numbers.

Second, many associations are reluctant to become involved in the new world of private finance if this leads to associations' having to charge higher rents and thereby becoming unable to perform their traditional role of meeting the needs of poorer households. In its response to the government consultation paper on finance for housing associations given before the 1988 Housing Act became law, the National Federation of Housing Associations (NFHA, 1987) expressed its worries that the boundaries between themselves and profit-making landlords would become blurred, that increased risk bearing would lead them to behave as purely commercial operators, and that rents would rise beyond the reach of the most needy.

After the passage of the 1988 Housing Act, it became clear that the NFHA's worries were not misplaced. Associations are becoming more commercially aware organizations—those that do not either cease expanding or risk insolvency (Kemp, 1989). At the time of writing, although some housing associations have keenly embraced the new world of private finance, risk bearing, and entrepreneurialism, many more remain cautious and skeptical of the supposed benefits that the new system will bring.

Third, and related to the last point, is the fact that even where associations are able to expand and increase the numbers of units for letting, a large proportion of this extra supply may go not to households in need, but to more "upmarket" demand groups, who can afford to pay higher rents. This may lead in the longer term to more units becoming available for low-income households through a process of filtering. However, even if this does occur, it is a "second-best" solution compared with the option of direct provision to poor households. Moreover, supply in general is not expanding fast enough to meet overall needs. Total housing output in the 1980s is below the peak levels of the 1960s and 1970s (Kleinman & Whitehead, 1988). Policy on improvement and repair has not kept pace with the rate of decline, while clearance and redevelopment have virtually come to an end (Gibson, 1986). Also, additional supply of middle-income housing may lead to increased demand from middle-income households, rather than initiating a chain of filtering moves.

DETERMINANTS OF VACANCY SUPPLY

Overall, then, the bulk of future supply to meet the need for social rental housing in the 1990s and beyond is likely to come from the local authority sector. Moreover, as we have seen, this supply will consist mainly of vacancies of the existing council stock.

Given that this is the case, the determinants of vacancy in the council sector are therefore of considerable importance. What are the factors that determine the level of vacancies generated from a particular stock? How do the present and future geographical patterns of vacancies—the major component of supply—correlate with indicators of housing need?

Vacancy rates vary considerably among authorities. In England, the average number of vacancies per 1,000 council dwellings between 1979-80 and 1986-87 was 38.9 per year. At the individual authority level, this varied from a low of 15.5 to a high of 106.2 (Eastall & Kleinman, 1989).

Using multiple regression analysis, we are able to identify the most important factors affecting the vacancy rate. The composition of the council sector in terms of the types of household residing there is the single most important determinant (Eastall & Kleinman, 1989). This reflects the importance of household dissolution in generating vacancies. That is, vacancies are higher in areas that contain a high proportion of elderly households and/or single-person households. Each of these household types is associated with a particular form of household dissolution (i.e., on death or on marriage) that leads to the creation of a vacancy.

Local housing market conditions, such as house prices and, to a lesser extent, private sector rents and private vacancies, are also found to be important. In areas of lower housing demand in the private sector, council tenants find it easier to move to owner-occupied or private rental housing, thereby creating vacancies in the council stock.

Hence the important determinants of the main source of the future rental supply—such as the profile of the existing tenants in terms of household type—bear no necessary relation to indicators of housing need. Indeed, the correlation between vacancies and housing need indicators is not particularly high. Moreover, factors such as high house prices both increase demand for council housing (from needy house-holds being priced out of the private sector) and reduce vacancies. Altogether, this suggests a growing geographical mismatch in the 1990s between the supply of and the need for cheaper rental housing.

CONCLUSIONS

In its first two terms (1979-83 and 1983-87), the Thatcher government's housing policy was essentially pragmatic, aimed primarily at the expansion of the owner-occupied sector. In Thatcher's third term, by contrast, there is evidence of a more thoroughgoing ideological approach that embraces both the owner-occupied and the rental sectors (Monk & Kleinman, 1989). This follows from the government's real-ization that there is an upper limit to the spread of home ownership; probably at least 25-30% of households will be tenants at any one time.

As far as rental housing is concerned, this approach envisages a change in the role of local authorities from direct providers of housing to "enablers" creating the conditions for other organizations to provide housing. In the government's view, the future supply of rental units will come from a dynamic independent private rental sector. However, this new course is likely to have only a minor impact at first, producing relatively small numbers of units even in the longer term and meeting the needs of middle- and higher-income rather than low-income groups.

Hence, in the 1990s, the main source of supply to meet housing need will be from the local authority sector. The supply of local authority vacancies in Britain, both new construction and vacancies in the exist-ing stock, has fallen considerably over the past ten years. This decline will continue into the 1990s. In its Expenditure Plans for the period 1989-90 to 1991-92, the government states that it "does not expect local authorities to add significantly to the stock of rental housing in their ownership" (H.M. Treasury, 1989, p. 8). Completions of dwellings by local authorities are planned to fall still further, from 15,000 per year in 1988-89 to only 6,000 annually in 1991-92. Meanwhile, sales of local authority dwellings will continue at a high level. These fell from a peak of 200,000 to 86,000 per year in 1986-87. However, in January 1987, the discount on sales of flats under the "right to buy" was increased, and sales rose to an estimated 150,000 in 1988-89. Sales have been projected to remain above 100,000 per year into the 1990s (H.M. Treasury, 1989, p. 10).

Hence the main source of supply of rental units to meet housing need in the 1990s will be vacancies in the existing council stock. This stock is not only shrinking in size, but also deteriorating in quality, with an estimated backlog of some £20 billion worth of repairs (Department of Environment, 1985).

Already there is some evidence of a divergence between where the supply of units is arising and where housing needs are greatest. The decline in supply over the past decade has been most severe in London, where access problems are most acute. In the future, this divergence is likely to widen, as the determinants of vacancy rates of supply bear no necessary relation to indicators of need.

This may well lead to a questioning of the established position that policy should aim at meeting needs where they arise. Already, there are some incentives in the new system of housing association finance for associations to build where costs are lowest, rather than where needs are greatest. In the future, those in housing need may be directed away from their present locations toward other parts of the country where housing pressures are less severe. Such a process would add the extra dimension of greater residential segregation to the increase in inequality that has been a key aspect of the "Thatcher decade."

NOTES

1. Housing associations are essentially private, but nonprofit organizations. They are run by voluntary committees, but normally employ professional staff. Mostly they are small in scale, catering to local needs, although there are a small number of much larger associations. For more details, see Hills (1987).

2. Demolition of existing dwellings is not a major factor in the decline in the size of the sector.

3. Both GNI and HNI have been criticized on the grounds of inadequately measuring housing need and of being susceptible to political pressures (see Bramley, 1989; Leather, 1984).

REFERENCES

Bover, O., Muellbauer, J., & Murphy, I. (1988). *Housing, wages and UK labour markets* (Discussion Paper No. 268). London: Centre for Economic Policy Research.

Bramley, G. (1989). *Meeting housing needs.* London: Association of District Councils.

Central Statistical Office. (1989). *Social trends no. 19.* London: HMSO.

Conway, J. (1984). *Capital decay.* London: SHAC.

Department of Environment. (1985). *An inquiry into the condition of the local authority housing stock in England: 1985.* London: Author.

Department of the Environment/Welsh Office. (1987). *Housing: The government's proposals* (Cmnd 214). London: HMSO.

Eastall, R., & Kleinman, M. P. (1989). A behavioural model of the supply of relets of council housing in England. *Urban Studies, 26,* 535-548.

Forrest, R., & Murie, A. (1988). *Selling the welfare state: The privatisation of public housing.* London: Routledge & Kegan Paul.

Gibson, M. (1986). Housing renewal: Privatisation and beyond. In P. Malpass (Ed.), *The housing crisis*. Beckenham: Croom Helm.

Hills, J. (1987). *When is a grant not a grant? The current system of housing association finance* (Welfare State Programme Discussion Paper No. 13). London: London School of Economics.

H.M. Treasury. (1989). *The government's expenditure plans 1989-90 to 1991-92* (Cmnd 609). London: HMSO.

Hughes, G. A., & McCormick, B. (1987). Housing markets, unemployment and labour market flexibility in the UK. *European Economic Review, 31*, 615-641.

Kemp, P. (1989). The demunicipalisation of rented housing. In M. Brenton & C. Ungerson (Eds.), *Social policy review 1988-9*. Harlow: Longman.

Kleinman, M. P. (1987). *Private finance for rented housing: An analytic framework* (Discussion Paper No. 18). Cambridge: University of Cambridge, Department of Land Economy.

Kleinman, M. P. (1988). Where did it hurt most? Uneven decline in the availability of council housing in England. *Policy and Politics, 16*(4), 221-235.

Kleinman, M. P., & Whitehead, C. M. E. (1988). British housing since 1979: Has the system changed? *Housing Studies, 3*, 3-19.

Leather, P. (1984). Housing allocations and the GNI. *Housing Review, 33*(1), 9-13.

London Research Centre. (1988). *Access to housing in London*. London: Author.

Minford, P., Peel, M., & Ashton, P. (1987). *The housing morass*. London: Institute of Economic Affairs.

Monk, S., & Kleinman, M. P. (1989). Housing. In P. Brown & R. Sparks (Eds.), *Beyond Thatcherism: Social policy, politics and society* (pp. 121-136). Milton Keynes: Open University Press.

National Federation of Housing Associations. (1987). *Rents, risks, rights*. London: Author.

Selbourne, D. (1987, May 18). Why Labour's attacks miss the mark. *Guardian*, p. 34.

Todd, J. E. (1986). *Recent private lettings 1982/84*. London: HMSO.

Whitehead, C. M. E., & Kleinman, M. P. (1986). *Private rented housing in the 1980s and 1990s*. Cambridge: Granta Editions.

Whitehead, C. M. E., & Kleinman, M. P. (1989). The private rented sector and the Housing Act, 1988. In M. Brenton & C. Ungerson (Eds.), *Social policy review 1988-9*. Harlow: Longman.

6

Recent Developments in Rental Housing in the United States

RICHARD P. APPELBAUM
PETER DREIER

THE UNITED STATES HAS no national housing program to assure the provision of adequate housing at affordable prices. Furthermore, a deeply rooted national belief in the sanctity of the "unfettered market-place" has an especially strong claim in the housing sector, which perhaps more than any other major economic arena is seen as embodying individual choice unrestrained by the hand of government.

Housing is symbolized by the free-standing single-family home, in which two out of every three people now live. Privately owned rental housing is widely viewed as being provided by "mom and pop" landlords who lease out a few rental units. The government enters the picture only as a last resort, providing subsidies to poor tenants or low-income housing lessees in hopes of restoring profitability to the private market in those instances where it fails.

The current regulatory framework has arisen over the past century in an entirely haphazard fashion. There is no national regulatory system, and localities have imposed restrictions according to prevailing political currents or local judicial decisions. Given the highly fragmented nature of the American system of local government, this has assured a patchwork of regulations and requirements that change over time and vary considerably within a single metropolitan area. Since local governments enjoy a large degree of autonomy in the United States, the federal government's role in directly regulating private housing has been minimal. The consequence of this crazy-quilt system has been to

97

impede local housing programs, subvert efforts at planning, and discourage even the consideration of a coherent federal policy.

Housing regulations are intended to satisfy two contrary objectives: assuring profitability and protecting the general safety and welfare. Zoning and land-use planning, for example, both regulate housing while serving to maintain property values. At the same time, safety, health, and building codes have produced a generally high-quality housing stock, while adding to costs and hence to prices. Such additional costs represent the price most Americans have been willing to pay for high-quality housing. The poor, however, cannot afford such trade-offs without government support. That support is now threatened.

At the present time the United States is experiencing a housing affordability crisis of major proportions. The number of low-rent units exceeded the number of low-income households by 2.4 million units (nearly one-third) in 1970; by the mid-1980s this had turned into a deficit of some 3.7 million units (Leonard, Dolbeare, & Lazere, 1989, p. 6). Needless to say, the shortfall was greatest among the poorest households, who could afford only the limited number of federally assisted units that were built during the period (approximately 5% of the total; see Clay, 1987, p. 4). Between 1978 and 1985, the number of low-rent units had declined by a half million; the number of low-income renters had grown by 3.6 million (Clay, 1987, p. 4). It is estimated that by 1985 there was a national shortage of some 3.3 million affordable units for households earning less than $5,000 annually—an increase of more than 80% since 1978 (Leonard et al., 1989, p. 9). This gap has been projected to rise to nearly 8 million units by the end of the century (Clay, 1987, p. 4), representing an estimated 18.7 million people (pp. i, 4, 24).

It is within this context that the present debates over "deregulation" must be understood: Government regulations are widely faulted for the rental housing crisis. The real estate industry in particular has long opposed government restraints on its activities. Their blaming of "excessive" regulation received especially strong support during the Reagan years, when "deregulation" and "privatization" were seen as the principal solutions to the housing crisis.

In the following pages we will briefly review the principal forms of housing tenure that exist in the United States, arguing that rental housing enjoys a second-class status that makes tenants particularly vulnerable both to market conditions and to the regulatory environment. We shall then look at the patchwork nature of the regulations that currently exist, before turning to the current attack on regulation. We shall argue that this attack ignores the key problem of how decent and affordable housing can be provided to low-income households, greatly exaggerating the costs of regulation. Finally, we shall conclude with some suggestions for government policy aimed at providing affordable *and* decent housing.

HOUSING TENURE
IN THE UNITED STATES

In a country where home ownership is regarded as a birthright, rental housing is the stepchild. With few exceptions, privately owned rental housing is seen as something to be endured until one can afford to make a down payment on a house (see Dreier, 1982b; Foley, 1980; Heskin, 1983; Michelson, 1977; Perrin, 1977; Rakoff, 1977). Permanent renting, in this prevailing view, is the unhappy fate of those too economically marginal to buy.[1] In contrast with many European and socialist countries, publicly assisted housing is thoroughly disdained as housing of last resort, the sole recourse of those who are too poor to afford even the least desirable private rental units.

Public policy and tax laws have long favored home ownership. In particular, federally guaranteed mortgages result in lowered interest rates, while at the same time private ownership confers significant tax advantages over renting: All mortgage debt service and local property taxes are fully deductible on state and federal tax returns. These factors have contributed to steady growth in home ownership, which peaked in 1980 at 65.5%.

Private rental housing has confronted a restricted market made up of those households for whom ownership is either undesirable or economically unviable: the elderly, unmarried urban professionals, and the poor. Rental housing in recent years has increasingly required subsidies, as higher-income renters are "creamed" off into home ownership; the "unsubsidized construction of new housing for low-income households is generally not economically feasible" (National Association of Home Builders, 1986, p. 16). As a result, most unsubsidized low-rent units were built before World War II.[2] Between 1974 and 1985, the number of privately owned, unsubsidized apartments renting for less than $300 per month (measured in 1988 dollars) fell by one-third, a loss of nearly 3 million units (Apgar, DiPasquale, McArdle, & Olson, 1989, p. 4). Downs (1983, pp. 100-109), in a model that simulated the investment profitably of rental units under various assumptions, found that rents in 1980 would have had to be 37% higher to sustain 1970 real values. At today's interest rates (approximately 10%), with slightly varying regional construction costs, the break-even point on newly constructed units is $700-$1,000 per month.

Government support is clearly needed to make low-cost housing economically viable. What form has such support taken? Given that housing is not an entitlement in the United States, the central thrust of federal housing policy has been to buttress private developers, builders, and apartment owners, under the belief that a properly functioning marketplace would solve all housing problems (see, for example, the articles in Bratt, Hartman, & Meyerson, 1986, Part II).

Assisted housing in the United States consists almost entirely of rental housing that is publicly subsidized, privately constructed, and owned (and managed) by either public housing authorities or private landlords. It currently reaches fewer than one out of every three poor households (Leonard et al., 1989, p. 27; U.S. Congressional Budget Office, 1988, p. 3).[3]

Since the 1940s, some 1.4 million units of public housing have been constructed, although new construction ceased under the latter years of the Reagan administration. Public housing construction is financed through federal grants to private developers, who then turn the units over to local housing authorities. Additional federal subsidies cover the difference between operating costs and 30% of tenant income.

Federal housing programs have in recent years attempted to entice private developers and landlords to build and/or rent low-income rental housing through a variety of mortgage, rental, and tax subsidy programs, administered primarily through Section 8 of the 1974 Housing Act. These programs have led to the construction of more than 2 million units of subsidized, privately owned rental housing since 1961 (National Low Income Housing Preservation Commission, 1988, p. 15).

The Reagan administration, in halting new construction of public and most private assisted housing, sought to substitute a system of vouchers and housing allowances that would enable tenants to find housing on the private market. The allowance partly fills the gap between what a poor family can afford and market rents. The problem is that the Reagan administration allocated only approximately one million certificates and vouchers for 1988 (U.S. Congressional Budget Office, 1988), while the need is at least six times as great as that. Furthermore, many housing markets are so tight that many families with vouchers cannot find housing in the private market.

When housing allowances are included, nearly 3 million renter households (as of 1987) receive some form of assistance in the private rental sector (U.S. Congressional Budget Office, 1988, Table 8; see also

National Low Income Housing Preservation Commission, 1988, p. 17). As the "affordability gap" (between tenants' incomes and the cost of providing housing) grows, the size of the subsidy required to make low-income housing profitable has necessarily increased, with the inflation-adjusted per unit cost of all federal programs growing by 70% between 1977 and 1987 (U.S. Congressional Budget Office, 1988, pp. 41, 43).[4] Needless to say, during fiscally austere times, rising per unit costs constitute a severe political liability, and predictably have contributed to recent cutbacks in housing expenditures (see below).

REGULATION OF RENTAL HOUSING

FEDERAL REGULATIONS

There is virtually no direct federal regulation of privately owned, unsubsidized rental housing. There are no federally mandated rent controls, protections against evictions, building code standards, or warranties of habitability.

Until recently, direct federal regulations were limited to units that received federal subsidies. Such restrictions were primarily intended to secure the economic viability of the federal government's stake in the unit, rather than to limit the private market in the interests of social objectives. For example, the initial 1935 and 1936 Federal Housing Administration (FHA) guidelines for federal housing insurance explicitly told underwriters to deny loans to neighborhoods experiencing such "adverse influences" as "the infiltration of inharmonious racial or nationality groups" (cited in Citizens' Commission on Civil Rights, 1986, p. 299).[5]

It was only in response to the civil rights movement of the 1960s that the federal government began to regulate private rental housing by guaranteeing access to minority groups. In 1962, the Kennedy administration issued the first fair housing orders, containing limited prohibitions against discrimination in federally assisted housing projects. Title VIII of the landmark 1968 Civil Rights Act contained broad prohibitions against discrimination in both private and public housing. Most recently, the Fair Housing Amendments Act of 1988 extended protections to the handicapped and families with children.

The principal impact of the federal government on private rental housing is not to be found in direct regulation, however, but rather in the indirect effects of credit and tax policies. Virtually all housing construction and purchase is financed through private credit in the United States, rendering housing costs and starts highly susceptible to fluctuations in credit availability and cost. Fluctuations in interest rates regulate the rental housing market in several ways. First, and most obvious, they have direct impact on rent levels via their effect on mortgage payments, which typically constitute at least half of landlords' total costs. Between the 1950s and the 1970s, interest rates nearly tripled; although they have moderated somewhat, they still stand at twice the earlier level. Interest rates have doubled since the 1950s; in the 1970s they nearly tripled. Second, interest rates also add directly to the cost of construction, sales prices, and other consumer credit. Construction loan interest is one of the most rapidly rising components of housing production costs (the other is land; see U.S. President's Commission on Housing, 1982, p. 181). Finally, interest rates have a strongly countercyclical effect on the housing economy, producing alternate periods of boom and bust. Federal income tax policies also serve to regulate rental housing indirectly, since they strongly affect profitability. Income property owners are able to reduce their taxable income not only by deducting actual expenses, but also by depreciating physical assets. This provides paper losses and therefore higher profits, while at the same time encouraging speculation, which drives up sales prices and thereby rents.

STATE AND LOCAL REGULATIONS

Local housing regulations have developed haphazardly, reflecting the often conflicting goals of the numerous jurisdictions that typically make up American metropolitan areas. The local regulatory environment reflects the tensions among health, safety, and environmental concerns, profit maximizing, and occasionally the organized voice of tenants. The degree to which regulations restrict or favor development can be understood only in terms of the balance of local political forces. While local politics are usually dominated by growth-oriented elites (Logan & Molotch, 1987; Molotch, 1976), in some locales they have recently confronted challenges both from organized tenants' movements and from home owner groups opposed to unrestricted development.

State and local housing regulation dates to New York City's Tenement House Act of 1867, which prescribed minimum standards for fire, safety, ventilation, and sanitation of multifamily housing, and the Tenement House Act of 1901, which created more adequate enforcement mechanisms (see Dreier, 1982b, p. 187; Lubove, 1962; Marcuse, 1986; Warner, 1972). The right of state and local governments to regulate housing has been upheld repeatedly by the courts; the right of localities to zone for land use dates to the landmark 1926 decision in *Village of Euclid v. Ambler Realty Co.* (272 U.S. 365). The Supreme Court's ruling in that decision reflects the central principal that a key function of local regulation is to maintain property values. The Court argued that zoning restrictions enhance constitutional property rights by providing a predictable environment for one's investment (Wheeler, 1988, p. 10). Regarding rental housing, the *Ambler* decision held that apartments could be legally restricted to prescribed zones because they posed a threat to the value of single-family homes.

Zoning proved to be a crude mechanism for controlling land use; boundaries were often arbitrary and—once drawn—legally inflexible, with the consequence that variances and nonconforming use permits were frequently granted on a case-by-case basis. As a result, cities and counties have devised a host of other means of regulating land use. While these have frequently been challenged in court as infringing on property rights, the courts have generally upheld the right to regulate so long as a "taking" of property does not occur. Such restrictions include health and safety requirements (generally a purely local matter) and minimal standards for building siting, design, and construction. Additionally, development in some communities is governed by locally enacted general plans to which particular projects must conform. Since the majority of states do not impose minimum standards on localities, the result is a mosaic of differing and often conflicting requirements that sometimes permit the construction of substandard housing.

Regulations managing growth and development (permits, impact fees, growth controls, linkage programs) have increased markedly in the past 20 years. While large-lot suburban zoning is not new, the effort to regulate urban growth through mandatory environmental review, down-zonings, service hookup moratoria, building permit allocation systems, and curbs on commercial growth represent relatively recent innovations. Growth controls are found primarily in the faster-growing regions of California, suburban New York and New Jersey, and Florida. They are frequently accompanied by regulations governing landscaping, environmental preservation, and architectural aesthetics, as well as special restrictions on land use in areas deemed to be of special historical, architectural, or environmental value. Finally, where municipal finances are limited,[6] efforts have been made to shift the costs of such infrastructure as roads and sewer lines from the general taxpayer to the developer, and thereby eventually to the final owner or renter.

In recent years, some communities have also adopted regulations aimed at increasing the supply of affordable housing. Among these are "inclusionary" zoning requirements, whereby developers of large-scale projects must include a designated number of units for specified low-income households or other designated categories, and "linkage" programs requiring commercial developers to help provide housing for the workers who will find eventual employment in their projects. Linkage programs have proven particularly attractive in "hot" commercial markets such as Boston and San Francisco (for useful discussions, see Dreier, Schwartz, & Greiner, 1988; Ehrlich & Dreier, 1990; Keating, 1986; Mallach, 1984).

The most visible efforts to address rising housing costs, however, consist of local rent controls. Since the early 1970s, tenant organizations have succeeded in enacting rent regulations in some 200 cities, primarily in Massachusetts, New York, New Jersey, and California (see Appelbaum & Gilderbloom, 1990; Dreier & Atlas, 1980, 1989; Gilderbloom & Appelbaum, 1988, chap. 7; Heskin, 1983).

THE ATTACK ON REGULATION

Beginning in the mid-1970s, big business went on the ideological offensive to change the public's perceptions of the profit system and the rightful role of government. Business leaders worried that the declining public confidence in big business revealed in public opinion polls would translate into antibusiness legislation. They viewed such progressive gains as the Equal Employment Opportunity Commission, the Occupational Health and Safety Administration, the Federal Trade Commission, and the Environmental Protection Agency as obstacles to corporate profits and a healthy economy.

This ideological offensive included major media and advertising campaigns, corporate financial support for conservative political candidates, and support for academic research by sympathetic scholars (see Dreier, 1982a). The triumph of this corporate campaign was the election of Ronald Reagan in 1980 on the promise to "get government off our backs." On a variety of fronts, the Reagan administration sought to dismantle progressive government policies that interfered with the unfettered property rights of the private sector. These efforts included a dramatic cut in tax rates for wealthy individuals and businesses; the deregulation (and often defunding) of agencies that protected consumers, workers, and the environment from corporate abuse; and sharp cutbacks in programs that provided the "safety net" of social benefits for low-income Americans. The thematic thread of this effort was embodied in the word *privatization.*

In this climate, the rental housing crisis of the late 1970s led conservative policymakers in both parties to identify "excessive" government interventions as the chief culprit, calling for an end to housing regulations altogether (Wheeler, 1988, pp. 4-5). While the call for privatization in the rental housing industry originated during the Democratic Carter presidency, it flourished during the fervently antiregulatory climate of the Republican Reagan administration.[7] In fact, President Reagan's Commission on Housing, composed primarily of lender and development interests, called for the dismantling of federal housing assistance and regulation. The increase in local rent and growth controls in the 1970s and 1980s in particular produced a backlash that echoed at the federal as well as state and local levels.

All forms of housing regulation are currently under attack. These include health, safety, and environmental regulations, growth controls and other limitations on development, and—especially—rent controls. The National Association of Home Builders (1986, p. 77), for example, faults "overregulation" for as much as 25% of home sales prices, while President Reagan's Commission on Housing (1982) has concluded that states should prohibit localities from limiting development unless "a vital and pressing governmental interest" can be demonstrated (p. 200). An important recent study commissioned for the bipartisan National Housing Task Force identified a large number of ways in which local land-use regulations are allegedly responsible for high building costs (Wheeler, 1988, p. 3);[8] impact fees and outdated building codes alone were held to contribute as much as 30% (p. 51). HUD Secretary Jack Kemp claims (without citing evidence) that regulations add $40,000 to $50,000 to the cost of a single-family home (Tufaro & Roth, 1989, p. 60), while the National Multi-Housing Council, a lobbying and information group that represents the largest owners and developers of rental housing in the United States, claims (also without citing evidence) that regulations currently add from $50 to $100 to monthly rents, thereby threatening the rental housing industry with extinction (Tufaro & Roth, 1989, pp. 60, 64).

Local rent controls have been held responsible for everything from rental housing shortages to homelessness. Although the courts have repeatedly upheld the legality of local rent regulations, their supposedly adverse effects have been used to justify federal and state efforts to ban or weaken local controls. Recent federal legislation appropriating funding for homeless programs required HUD to undertake a study of the effects of rent control, in order to determine whether or not rent control actually worsened housing problems (including homelessness).

THE ATTACK ON FEDERAL REGULATIONS

Despite the general retreat on civil rights during the Reagan administration, HUD continued to pursue fair housing legislation; prohibitions against racial discrimination are relatively costless. Moreover, federal antidiscrimination laws in the field of housing are poorly enforced (Citizens' Commission on Civil Rights, 1986, pp. 309-310); after 20 years, racial discrimination remains pervasive. The recently enacted 1988 Fair Housing Amendments Act does in fact impose some cost burdens on the private sector by requiring handicap accessibility, but these costs are relatively minor determinants of overall housing affordability.

Credit and tax policies during the Reagan era tended to disfavor rental housing construction strongly. Tight money policies (intended to curb inflation) raised the real cost of credit, once the rate of inflation fell significantly below mortgage interest rates. More significantly, the 1986 Tax Reform Act removed a number of major incentives for investment in rental housing. It contained deep cuts in tax rates for upper-income taxpayers, repealed accelerated depreciation, and disallowed so-called passive housing investments (sheltering income through the ownership of rental property in which one has no direct managerial involvement). The loss of such tax advantages reduced the value of rental property by an estimated one-fifth (Furlong, 1986, p. 16). This, in turn, is expected to force compensating average rent increases of up to one-quarter over the rate of inflation by 1991 (see Apgar, Brown, Doud, & Schink, 1985, p. 1; Apgar & Doud, n.d., p. 2). The National Association of Home Builders predicts that rental housing construction will decline by half as a direct result (Furlong, 1986, p. 16), while an MIT market simulation predicts that some 1.4 million fewer units will be constructed by 1994 than would have occurred in the absence of tax reform (Apgar et al., 1985, p. 1).

The Reagan administration's principal efforts at deregulation were directed at drastically reducing the already limited federal role in housing. HUD's budget for subsidized housing was cut by four-fifths, from $33 billion to less than $8 billion, eviscerating publicly owned and subsidized housing. Moreover, the Reagan administration—following Thatcher's lead in England—implemented a demonstration program to sell off public housing units, in hopes of showing the feasibility of more widespread privatization (see Dreier, 1986). This plan was originally proposed by Jack Kemp, at the time a member of Congress. Now President Bush's secretary of HUD, Kemp remains a staunch advocate of privatization of public housing. The greatest impetus for privatization of the existing federally assisted housing stock will not come from actions aimed at deregulation, however, but rather from simple government inaction. Many existing federal subsidies are likely to expire within the next 10 years, returning a large portion of the federally assisted housing stock to the open marketplace. According to Leonard et al. (1989, chap. 5), nearly a million Section 8 certificates and voucher contracts will expire by 1995, including some 300,000 as early as 1991. Additionally, some 368,000 subsidized units will be eligible for mortgage prepayments during the next 15 years. One study estimates that unless federal action is taken, prepayments will be made on two-thirds of these, allowing them to raise their rents to market levels (National Low Income Housing Preservation Commission, 1988). Finally, some 280,000 subsidized units will require substantial repair over the next 15 years if they are to avoid default. Unless the Bush administration is willing to commit nearly $90 billion for these various purposes over the next decade, this privatization of much of the existing supply of federally assisted housing will result in its permanent loss to low-income households.

THE ATTACK ON
STATE AND LOCAL REGULATIONS

Recent court decisions have somewhat limited the right of local jurisdictions to enact zoning and land-use regulations. In *Nollan v. California Coastal Commission* (1987, 55 L.W. 5145), for example, the Supreme Court overruled an exaction (imposed on a home owner) on the grounds that the fee had not been shown to relate clearly to the proposed development. While this was widely interpreted as a limitation on exactions in general, in fact it upholds their legality when their connection to a proposed project can be demonstrated (Wheeler, 1988, p. 15). More significantly, in *First English Evangelical Lutheran Church of Glendale v. County of Los Angeles* (1987, 55 L.W. 4782), the Court ruled that owners who have been restricted from developing their property be compensated for any losses if the restriction is subsequently overturned as unlawful.[9] This may well have a chilling effect on the willingness of localities to regulate, particularly in light of *Nollan* (Wheeler, 1988, p. 17).

There has been a great deal of research on the impact of regulations on housing costs, much of it consisting of anecdotes about particular projects or statistical analyses that fail to compare locales with varying degrees of regulation. Regulations that add to developers' costs have been estimated to add anywhere from 2% to 30%, with the higher figures generally coming from industry reports (see, for example, Bickert, Browne, Coddington, & Associates, 1976; California Construction Industry Research Board, 1975; Franklin & Muller, 1977; Frieden, 1979, 1982; Gruen & Gruen, 1977; Katz & Rosen, 1980; National Association of Home Builders, 1982; Orange County Cost of Housing Committee, 1975; Stull, 1974; Tagge, 1976). Despite the belief that the U.S. housing economy is "overregulated," most research indicates that local regulations are relatively minor determinants of housing production and costs. One study, for example, found that 71% of national housing starts are predicted simply by the growth of constant-dollar GNP (Weintraub, 1982, pp. 89-90). Another found that per capita income along with solar radiation and seasonal temperature differences (both seen as indicators of regional demand shifts) explained almost two-thirds of the variance in housing prices (Stutz & Kartman, 1982, p. 232). In most housing markets, new construction typically accounts for only a small percentage of the total housing stock, and restrictions on construction are likely to have only a marginal effect on overall price levels. One study has concluded that "production of new housing in most cities amounts to no more than 3 percent of the existing stock and thus a 50 percent cut in this *flow* supply would amount to a cut of only 1.5 percent in *stock* supply in the first year" (Markusen, 1979, p. 153).

An extensive review of a large number of studies concerned with the relationship between various forms of regulation and housing prices concludes that *supply restrictions* typically add less than 10% to housing, generally in the 5% range (Appelbaum, 1985; Gilderbloom & Appelbaum, 1988, chap. 6). One econometric analysis concludes that "the combined effect of increasing development densities by one unit per acre, reducing development fees by 50 percent, and doubling supplies of vacant land—all drastic steps—would be to lower the sales price of a new home by . . . roughly 6 percent" (Dowall & Landis, 1982, p. 88). Direct regulatory costs have been characterized as especially onerous, yet most research indicates that they generally add at most a few percent to total costs. The effect of rent controls on investment in rental housing has been studied extensively; prevailing economic theory notwithstanding, virtually all studies to date have concluded that moderate rent controls have no adverse effects on either new construction or maintenance of controlled units.[10]

THE FUTURE OF REGULATION

The regulation of U.S. housing can move in several directions, depending on the political coalitions that emerge over the next decade.

First, the continued ascendancy of the New Right/Republican agenda could continue the current climate of privatization. Much depends on whether the business community—with its strong influence on the Republican party—makes housing issues a priority on its agenda. The HUD scandal[11] has reinforced public conceptions of government housing programs as wasteful and mismanaged, making it more difficult for cities to enforce local regulations; some have even called for the abolition of HUD (see "Abolish HUD," 1989). The Reagan appointments to the Supreme Court are likely to be influential for several decades, providing judicial support for the assault on government intervention in private property.

Second, the traditional liberal agenda continues to emphasize support of the for-profit sector, through modest expansion of low- and moderate-income housing, preservation of the existing assisted housing stock, and continued provision of housing allowances for poor families occupying private rental units. The Reagan-era delegitimation of government intervention has clearly put the liberal agenda on the defensive; even vocal congressional liberals have a difficult time simply calling for "more of the same" in the current climate. The 1988 National Housing Task Force, whose recommendations are embodied in the major Democratic-sponsored housing bills currently pending in the Congress, represented an attempt to re-create the old "housing coalition" between real estate and the urban poor; its measures call for only minor additions to current spending levels. The bills would not even come close to restoring the federal government to housing assistance levels of the pre-Reagan years.

Finally, a small but growing sector of progressive housing advocates and developers is advocating a new direction for American housing policy based on the social democratic approaches popular in Canada, Western Europe, and Sweden. This approach, currently embodied in at least three legislative measures,[12] is supported by low-income housing advocates, community development corporations, and other nonprofit developers. It seeks to remove housing increasingly from the speculative market, transforming it into limited-equity, resident-controlled housing that is funded through direct capital grants rather than long-term debt.

In the present political climate, it appears unlikely that substantial federal dollars will be spent on innovative approaches to nonmarket housing. More likely, progressives will have to be content with extremely limited funding for nonprofit housing, while liberals struggle to revive their traditional market-oriented programs in the face of conservative resistance to direct subsidies for low-cost housing. The latter will continue to press for privatizing the remaining public and assisted housing stock, while attempting to force localities to further curb housing regulations (particularly rent controls). Unfortunately, the federal budget deficit, Gramm-Rudman deficit limits, and the unwillingness of Congress to propose major cuts in military spending or to increase taxes mean that badly needed investment in housing will continue to compete with child care, health care, and education.

NOTES

1. Renters are on the average poorer than home owners in the United States, and are becoming more so: While median renter income stood at 60% of median home owner income in 1967, by 1977 the ratio had dropped to 55%, and by 1987 to 48% (Apgar, 1988, exhibit 14).

2. There has been a growing gap between tenants' incomes and the costs of producing and renting apartments. In 1986, one out of every four full-time jobs did not pay sufficient income to lift a four-person family out of poverty, while the median income of all renters stood at only 85% of 1972 levels (National Housing Task Force, 1988, p. 5).

3. Only "very low-income" households (less than half of median local income) are currently eligible for federally assisted housing; prior to 1981 "low-income" households (50-80% of median local income) were also eligible. Some 6-7 million low-income renter families receive no housing assistance whatsoever. (For summaries and discussions of these programs, see Bratt, 1986; National Association of Home Builders, 1986, pp. 44-47; National Low Income Housing Preservation Commission, 1988, chap. 1; U.S. Congressional Budget Office, 1988, chap. III.)

4. This is despite the fact that the tenant's contribution was raised from 25% to 30% of household income during the same period.

5. The quote is taken from *The Richmond School Decision: Complete Text of Bradley vs. School Board of Richmond* (1972, p. 172). Such restrictive covenants were ruled unenforceable by the Supreme Court in *Shelley v. Kraemer* in 1948.

6. This has sometimes occurred in otherwise prosperous cities as a result of "taxpayer revolt" measures limiting property taxes and local government spending, such as California's Proposition 13 (which cut back property taxes by two-thirds) and the Gann Amendment (which placed limits on increases in state and local spending in California).

7. For example, see President Carter's Commission on Housing (1979), as well as other federal studies of the housing crisis conducted during his administration (e.g., U.S. Comptroller General, 1979; U.S. Department of Housing and Urban Development, 1979).

8. These include raising the cost of land through large-lot zoning or restricting the supply of land, promoting inefficient use of space through restrictions against multifamily housing, establishing time-consuming regulatory bureaucracies, engendering citizen-sponsored legal challenges to development, and reducing competition by discouraging builders with multiple restrictions.

9. Previously, no compensation was provided if an ordinance was legally invalidated; the affected owners could simply proceed with their original development plans (Wheeler, 1988, p. 16).

10. For reviews of dozens of existing studies as well as an empirical analysis of rent control in New Jersey, see Gilderbloom and Appelbaum (1988, chap. 7) and Appelbaum

and Gilderbloom (1990); for an industry-sponsored review that draws opposite conclusions, see Downs (1988). The claim that rent control causes homelessness (by discouraging the production of rental housing) can be traced to a single widely published study (Tucker, 1987a, 1987b, 1989a, 1989b) that has been shown to be methodologically flawed; see Appelbaum, Dolny, Dreier, and Gilderbloom (1989) and Quigley (1989) for critiques.

11. The agency has since been revealed to be thoroughly scandal-ridden during the Reagan period, with its limited resources going in large part to private developers who contributed heavily to Republican candidates as well as lobbyists and political cronies. This has contributed to the call to dismantle HUD altogether (see "Abolish HUD," 1989). See Dreier (1989) and also Chapter 9 of this volume.

12. The most far reaching of these is Representative Ron Dellums's National Comprehensive Housing Act, which calls for annual federal expenditures of roughly $50 billion in direct capital grants to both nonprofits and public housing authorities. The money would be used primarily to build and rehabilitate low-cost housing, as well as to purchase existing privately owned housing for transfer to nonprofit organizations. All housing constructed, rehabilitated, or purchased under the Dellums bill would remain in the "social" sector, never to be burdened with debt. Occupants would pay only the operating costs, which would dramatically lower their rents (for further information, see Institute for Policy Studies, 1989). The other measures are scaled-down versions of parts of the Dellums bill; these include Representative Barney Frank's National Affordable Housing Act ($15 billion) and Representative Joseph Kennedy and Senator Frank Lautenberg's Community Housing Partnership Act ($500 million).

REFERENCES

Abolish HUD. (1989, August 21). *New Republic*, pp. 7-8.

Apgar, W. C. (1988). *The nation's housing: A review of past trends and future prospects for housing in America* (MIT Housing Policy Project HP1). Cambridge: MIT Center for Real Estate Development.

Apgar, W. C., Brown, H. J., Doud, A. A., & Schink, G. A. (1985). *Assessment of the likely impacts of the president's tax proposals on rental housing markets.* Cambridge: Joint Center for Housing Studies of MIT and Harvard University.

Apgar, W. C., DiPasquale, D., McArdle, N., & Olson, J. (1989). *The state of the nation's housing, 1989.* Cambridge: Joint Center for Housing Studies of MIT and Harvard University.

Apgar, W. C., & Doud, A. A. (n.d.). *Tax reform and national housing policy.* Philadelphia: Arthur Doud Associates.

Appelbaum, R. P. (1985). *Regulation and the Santa Barbara housing market.* Berkeley: California Policy Seminar.

Appelbaum, R. P. (1989, May-June). The affordability gap. *Society*, pp. 6-8.

Appelbaum, R. P., Dolny, M., Dreier, P., & Gilderbloom, J. I. (1989). *Scapegoating rent control: Masking the causes of homelessness.* Washington, DC: Economic Policy Institute.

Appelbaum, R. P., & Gilderbloom, J. I. (1990). Rental housing under modern rent control. *Environment and Planning.*

Bickert, Browne, Coddington, & Associates. (1976). *An analysis of the impact of state and local government intervention on the homebuilding process in Colorado, 1970-1975.* Denver: Association of Housing and Building.

Bratt, R. G. (1986). Public housing: The controversy and the contribution. In R. G. Bratt, C. Hartman, & A. Meyerson (Eds.), *Critical perspectives on housing*. Philadelphia: Temple University Press.

Bratt, R. G., Hartman, C., & Meyerson, A. (Eds.). (1986). *Critical perspectives on housing*. Philadelphia: Temple University Press.

California Construction Industry Research Board. (1975). *Cost of delay prior to construction*. Los Angeles: Author.

Checkoway, B. (1980). Large builders, federal housing programmes, and postwar suburbanization. *International Journal of Urban and Regional Research, 4*, 21-45.

Children's Defense Fund. (1988). *Vanishing dreams: The growing economic plight of America's young families*. Washington, DC: Author.

Citizens' Commission on Civil Rights. (1986). The federal government and equal housing opportunity: A continuing failure.In R. G. Bratt, C. Hartman, & A. Meyerson (Eds.), *Critical perspectives on housing*. Philadelphia: Temple University Press.

Clay, P. I. (1987). *At risk of loss: The endangered future of low-income rental housing resources*. Washington, DC: Neighborhood Reinvestment Corporation.

Dowall, D. E., & Landis, J. (1982). *Land-use controls and housing costs: An examination of San Francisco-Bay Area communities* (Reprint no. 196). Berkeley: University of California, Institute of Urban and Regional Development.

Downs, A. (1983). *Rental housing in the 1980s*. Washington, DC: Brookings Institution.

Downs, A. (1988). *Residential rent controls: An evaluation*. Washington, DC: Urban Land Institute.

Dreier, P. (1982a). Capitalists vs. the media: An analysis of an ideological mobilization among business leaders. *Media, Culture, and Society, 4*(2), 111-132.

Dreier, P. (1982b). The status of tenants in the United States. *Social Problems, 30*(2), 179-198.

Dreier, P. (1986, August 4). Private project: Public housing for sale. *New Republic*, pp. 13-15.

Dreier, P. (1989, August 21-28). Communities, not carpetbaggers: Lessons of the HUD scandal. *The Nation*, pp. 198-202.

Dreier, P., & Atlas, J. (1980, January). The housing crisis and the tenants' revolt. *Social Policy*, pp. 13-24.

Dreier, P., & Atlas, J. (1989, Winter). Grassroots strategies for the housing crisis: A national agenda. *Social Policy*, pp. 25-38.

Dreier, P., Schwartz, D. C., & Greiner, A. (1988, September-October). What every business can do about housing. *Harvard Business Review*, pp. 52-61.

Ehrlich, B., & Dreier, P. (1990). Downtown development and urban reform: The politics of Boston's linkage policy. *Urban Affairs Quarterly, 25.*

Foley, D. L. (1980). The sociology of housing. In A. Inkeles et al. (Eds.), *Annual review of sociology* (Vol. 6, pp. 457-478). Palo Alto, CA: Annual Reviews.

Franklin, J. J., & Muller, T. (1977). Environmental impact evaluation, land use planning, and the housing consumer. *Journal of the American Real Estate and Urban Economics Association, 5*(3), 279-301.

Frieden, B. J. (1979). *The environmental protection hustle*. Cambridge: MIT Press.

Frieden, B. J. (1982). The exclusionary effect of growth controls. In M. B. Johnson (Ed.), *Resolving the housing crisis*. San Francisco: Pacific Institute for Public Policy Research.

Furlong, T. (1986, August 20). Real estate industry finds changes devastating. *Los Angeles Times*, Part I, p. 17.

Gilderbloom, J., & Appelbaum, R. P. (1988). *Rethinking rental housing.* Philadelphia: Temple University Press.

Gruen, C., & Gruen, N. (1977). *Rent control in New Jersey: The beginnings.* Sacramento: California Housing Council.

Hartman, C. (1975). *Housing and social policy.* Englewood Cliffs, NJ: Prentice-Hall.

Heskin, A. D. (1983). *Tenants and the American dream.* New York: Praeger.

Institute for Policy Studies. (1989). *The right to housing: A blueprint for housing the nation.* Oakland, CA: Community Economics.

Katz, L., & Rosen, K. T. (1980). *The effects of land-use controls on housing prices* (Working Paper no. 80-13). Berkeley: University of California, Center for Real Estate and Urban Economics.

Keating, W. D. (1986, Spring). Linking downtown development to broader community goals. *American Planning Association Journal.*

Leonard, P. A., Dolbeare, C. N., & Lazere, E. B. (1989). *A place to call home: The crisis in housing for the poor.* Washington, DC: Center of Budget and Policy Priorities, Low Income Housing Information Service.

Logan, J., & Molotch, H. (1987). *Urban fortunes: The political economy of place.* Berkeley: University of California Press.

Low Income Housing Information Service. (1989). *The 1990 low income housing budget.* Washington, DC: Author.

Lubove, R. (1962). *The progressives and the slums.* Pittsburgh: University of Pittsburgh Press.

Mallach, A. (1984). *Inclusionary housing programs: Policies and practices.* New Brunswick, NJ: Rutgers University, Center for Urban Policy Research.

Marcuse, P. (1986). Housing policy and the myth of the benevolent state. In R. G. Bratt, C. Hartman, & A. Meyerson (Eds.), *Critical perspectives on housing.* Philadelphia: Temple University Press.

Markusen, J. R. (1979). Elements of real asset pricing: A theoretical analysis with special reference to urban land prices. *Land Economics, 55*(2), 153-166.

Michelson, W. (1977). *Environmental choice, human behavior and residential satisfaction.* New York: Oxford University Press.

Mollenkopf, J. (1975). The postwar politics of urban development. *Politics and Society, 5,* 247-295.

Molotch, H. (1976). The city as a growth machine: Toward a political economy of place. *American Journal of Sociology, 82*(2), 309-332.

National Association of Home Builders. (1982). *Land use and construction regulations: A builders' view.* Washington, DC: Author.

National Association of Home Builders. (1986). *Low- and moderate-income housing: Progress, problems, and prospects.* Washington, DC: Author.

National Housing Task Force (Rouse-Maxwell Commission). (1988). *A decent place to live.* Washington, DC: Author.

National Low Income Housing Preservation Commission. (1988). *Preventing the disappearance of low income housing.* Washington, DC: National Corporation for Housing Partnerships.

Orange County Cost of Housing Committee. (1975). *The cost of housing in Orange County.* Orange, CA: Author.

Perrin, C. (1977). *Everything in its place.* Princeton, NJ: Princeton University Press.

Quigley, J. (1989). *Does rent control cause homelessness? Taking the claim seriously* (Working Paper no. 166). Berkeley: University of California, Graduate School of Public Policy.

Rakoff, R. M. (1977). Ideology in everyday life: The meaning of the house. *Politics and Society, 7,* 85-104.

Stull, W. J. (1974). Land use and zoning in an urban economy. *American Economic Review, 64*(3), 337-347.

Stutz, F. P., & Kartman, A. E. (1982). Housing affordability and spatial price variations in the United States. *Economic Geography, 58*(3), 221-235.

Tagge, C. (1976). *The effects of public regulation on the costs of new single family units.* Unpublished master's thesis, University of Texas, Austin.

Tucker, W. (1987a, September 25). Where do the homeless come from? *National Review,* pp. 32-43.

Tucker, W. (1987b, November 20). Where do the homeless come from? Manhattan Institute for Policy Research *Associates Memo, 5.*

Tucker, W. (1989a, January 12). America's homeless: Victims of rent control. *Heritage Foundation Backgrounder, 685.*

Tucker, W. (1989b). Where do the homeless come from? In J. Elliot (Ed.), *Urban society* (4th ed., pp. 111-114). Guilford, CT: Dushkin.

Tufaro, D. F., & Roth, R. R. (1989, July). [Article]. *National Real Estate Investor,* pp. 60-64.

U.S. Bureau of the Census. (1981). *Annual housing survey: 1980, Part C, financial characteristics of the housing inventory, U.S. and regions* (Current Housing Reports, Series H-15—80). Washington, DC: Government Printing Office.

U.S. Comptroller General. (1979). *Rental housing: A national problem that needs immediate attention.* Washington, DC: General Accounting Office.

U.S. Congressional Budget Office. (1988). *Current housing problems and possible federal responses.* Washington, DC: Government Printing Office.

U.S. Department of Housing and Urban Development. (1979). *Final report of the Task Force on Housing Costs.* Washington, DC: Government Printing Office.

U.S. President's Commission on Housing. (1979). *Report.* Washington, DC: Government Printing Office.

U.S. President's Commission on Housing. (1981). *Interim report.* Washington, DC: Government Printing Office.

U.S. President's Commission on Housing. (1982). *Report of the President's Commission on Housing.* Washington, DC: Government Printing Office.

U.S. President's Commission on Housing. (1983). *Final report.* Washington, DC: Government Printing Office.

Warner, S. B. (1972). *The urban wilderness.* New York: Harper & Row.

Weintraub, R. E. (1982). Private housing starts and the growth of the money supply. In M. B. Johnson (Ed.), *Resolving the housing crisis.* San Francisco: Pacific Institute for Public Policy Research.

Wheeler, M. (1988). *Resolving local regulatory disputes and building consensus for affordable housing* (MIT Housing Policy Project HP9). Cambridge: MIT Center for Real Estate Development.

Part III

Public Housing

Introduction

RACHEL G. BRATT

THE INTERPLAY BETWEEN public and private interests in the U.S. public housing program has been present ever since its inception in the late 1930s. Although the following overview is specific to the American experience, other Western countries often have initiated identical programs as their own public housing programs have evolved.

At least five distinct phases of private sector involvement with the U.S. public housing program can be identified:

(1) private sector opposition to public housing legislation (1937)
(2) creation of two variants of public housing: leased and turnkey (1965)
(3) alternatives to public housing: introduction of new housing subsidy programs that operated through the private sector (1959-1974)
(4) transfer of management functions to tenants (1970s-1980s)
(5) sales of public housing (1980s)

The chapters that follow address the last two of these phases. Silver addresses the sale of public housing in the United States and Werczberger addresses tenant management of public housing in Israel. Following a brief look at the first three phases outlined above, we will return to these two issues.

PRIVATE SECTOR OPPOSITION TO THE ORIGINAL LEGISLATION, 1937

In the United States, the private sector had a major role in shaping the public housing program, although primarily in a negative way. The views generally held by the private home-building industry were summarized by the president of the National Association of Real Estate

Boards: "Housing should remain a matter of private enterprise and private ownership. It is contrary to the American people and the ideals they have established that government become landlord to its citizens" (quoted in Keith, 1973, p. 33).

Conservatives in the U.S. Congress labeled public housing a socialist program on the grounds that government should not be in competition with the interests of private entrepreneurs (Friedman, 1968; Keith, 1973). Under pressure from the housing industry, the public housing legislation included an "equivalent elimination" provision that required local housing authorities to eliminate a substandard or unsafe dwelling unit for each new unit of public housing built. Public housing could replace inadequate units, but it was not to increase the overall supply, since that could dampen rents in the private market. The effectively waged argument that public housing should not compete with privately owned housing had an important outcome: Its stereotypically austere physical design is often easily recognized as public housing, standing apart from the overall housing stock (see Bratt, 1989).

After a disruption in public housing production during most of the 1940s, due to World War II, public housing squeaked back onto the federal agenda in 1949. Between 1949 and 1959, nearly 50% more public housing units were built compared with the prior decade. During that period public housing provided homes to predominantly White, "working poor" households, but by the 1960s the public housing population became predominantly very low income and non-White. As the public housing stock grew older, operating and repair costs rose, and shortfalls could not be made up from tenants' rents. Subsidies to fill the gap were eventually provided, but funding delays often meant that buildings deteriorated and living conditions declined. Much of the poor image of public housing can be traced to the lack of adequate operating funds and to its often easily distinguished appearance. At least some of the blame for these dual problems lies with the relentless insistence of the private sector that government spend as little as possible and make certain that public housing would be decidedly different from the private housing stock.

CREATION OF TWO VARIANTS OF PUBLIC HOUSING, LEASED AND TURNKEY, 1965

Despite the largely negative role of the private sector in the early years of the public housing program, when problems in the public housing program captured attention during the 1960s the private sector was looked to as the savior. In 1965, two variations on the public housing program were created, both of which included central roles for the real estate industry. First, the Section 23 leased housing program authorized local housing authorities to enter into long-term leases with private owners of apartments. Low-income tenants were then selected by local housing authorities; the former paid 25% of their income in rent and the latter paid the difference between this amount and the contracted rent.

Second, "turnkey" public housing enabled private developers to contract with local housing authorities to construct projects. Upon completion of a project, the developer sold it and "turned the key" over to the housing authority at a predetermined price.

Eugene Meehan (1979) has questioned why this shift of control from the public to the private sector was expected to lead to significant improvements in the public housing program:

> Some proponents of privatization simply opposed public ownership on ideological grounds and took the "failure" of public housing as evidence favorable to their position. Others accepted the popular belief in the superiority of private enterprise (as opposed to the intrinsic evil of public ownership). In most cases, supporters of privatization apparently had only the vaguest notion of how and why it would improve the housing program. (p. 136)

According to Meehan, privatization was a costly error, and

> the assumption that privatization was justified because public ownership had been tried and found wanting was grossly mistaken. . . . The poor quality and performance of some of the controversial developments were no more a simple function of public ownership than the poor performance and quality of some recent turnkey developments were a simple function of private development. (pp. 205-206)

ALTERNATIVES TO PUBLIC HOUSING: NEW HOUSING SUBSIDY PROGRAMS THAT OPERATED THROUGH THE PRIVATE SECTOR, 1959-1974

Although not directed at public housing itself, a series of housing subsidy programs were initiated by the federal government, starting in 1959, that involved sponsorship first by private nonprofit and later by private for-profit limited-dividend developers. These new programs effectively altered thinking about how housing for low- and moderate-income people should be subsidized and provided concrete alternatives to the public ownership and management approach of the public housing program.

The Section 202 program, authorized in 1959, provided low-interest loans to nonprofit sponsors of housing for the elderly. Two years later, the Section 221(d)(3) program provided low-interest loans to both nonprofit and for-profit developers of housing for low- and moderate-income households. In 1968, the Section 235 and 236 programs subsidized the interest rate on loans to as low as 1% to low- and moderate-income home buyers and to private landlords agreeing to rent their units to income-eligible occupants, respectively.

During the 1960s, enthusiasm for this public-private approach to low-income housing grew. Private interest groups became forceful lobbyists for publicly subsidized, privately owned housing. President Johnson's Committee on Urban Housing (1968), which was composed primarily of representatives from the private home-building industry, fully endorsed the public-private approach, while providing only slightly veiled criticism of the public housing program:

> The nation has been slow to realize that private industry in many cases is an efficient vehicle for achieving social goals. . . . some programs still make too little use of the talents of private entrepreneurs. . . . One of the basic lessons of the history of Federal housing programs seems to be that the programs which work best . . . are those that channel the forces of existing economic institutions into productive areas. This approach has proved to be better than wholly ignoring existing institutions and starting afresh outside the prevailing market system. Reliance on market forces should be increased in the future. (p. 54)

In 1974, the home-building industry was successful in its lobbying efforts. In addition to the creation of the Section 8 Existing Housing program, which, similar to the leased housing program, provided a rental subsidy to eligible tenants living in private rental units, Congress enacted the Section 8 New Construction/Substantial Rehabilitation program to stimulate the production of affordable housing by private nonprofit and for-profit developers.

After two decades of experimentation with programs that provide a public subsidy to private owners of rental housing, it is now clear that there are serious drawbacks in this approach's ability to provide long-term affordable housing. Although hundreds of thousands of people have been provided with housing, the goals of private, for-profit developers often conflict with those of low-income residents and with the public's need to maintain a good, affordable stock of low-cost shelter. Expiring use restrictions, whereby private owners of subsidized housing are allowed to convert their buildings to high-cost rental units or to sell them as condominiums at the end of their regulatory agreements with HUD, provide a good example of how profit-maximization goals and public goals do not easily coexist (see Chapters 6 and 9, this volume).

TRANSFER OF MANAGEMENT FUNCTIONS TO TENANTS, 1970s-1980s

Conditions in many public housing developments worsened during the 1960s and 1970s. As tenants began to protest the poor conditions, the idea of tenant management of public housing gained credibility in many cities. One of the best-known examples of tenant-managed public housing is in St. Louis, where tenant management corporations (TMCs) oversee the operation of more than 3,000 apartments in five family developments.

Based on experiences in St. Louis, as well as similar programs in several other cities, HUD, with the assistance of the Ford Foundation, launched a three-year demonstration program in 1976. Among the positive results were that "residents were able to manage their developments as well as prior management had and, in so doing, [were able] to provide employment for some tenants and increase the overall satisfaction of the general resident population" (Manpower Demonstration Research Corporation, 1981, p. 239). Further positive findings were that tenant management was continued even after HUD's supplemental funds were exhausted, operations continued to improve, and, in subsequent years, additional TMCs were formed (Kolodny, 1986). Although the same report also notes that several TMCs were not successful and questions the additional costs of tenant management, tenant management is thought to have significant potential.

The Housing and Community Development Act of 1987 gave credibility to tenant management by authorizing direct funding for the development of TMCs. HUD was allocated some $5 million in fiscal years 1988 and 1989 for this purpose. While a step that clearly endorses the concept of tenant management, it is of such a small scale that tenant management will not, at least at the present time, become a dominant feature of local housing authorities' operations.

In Chapter 8, Werczberger presents the results of a tenant management program launched in Israel in 1984. The key question is whether the program was successful in changing "residents' attitudes toward maintenance by inducing them to organize and actively participate in the renovation of their buildings." Paralleling some of the results of the U.S. demonstration discussed above, Werczberger reports that residents of the experimental buildings "are clearly more satisfied with their dwellings and, to some extent, also with their buildings than those living in the control groups." Yet differences in maintenance quality generally were not found, and "participation in the program did not lead to the radical change in attitudes and behavior the initiators had hoped for."

While tenant management of public housing should continue to be explored, privatizing this function may not always be in the tenants' best interests. If public housing management were exemplary, the need for residents to expend time and energy on the upkeep of their living environment would be unnecessary. Instead, they would be able to spend more time on work, family, leisure, and educational activities. It is legitimate to question why tenants in publicly subsidized housing should do more maintenance than tenants in unsubsidized dwellings. All tenants pay (although some at a subsidized rate) for the "luxury" of their housing maintenance being taken care of by the landlord, whether public or private.

SALES OF PUBLIC HOUSING, 1980s

When privatization is discussed with reference to public housing, it is the sale of public housing units that most readily comes to mind. Following the public housing sales program initiated in Great Britain under Prime Minister Margaret Thatcher, the Reagan administration, as well as several members of Congress (among them Jack Kemp, currently secretary of HUD) became attracted to the idea. But, as Silver points out in Chapter 7, conservatives were also joined by liberal democrats, who saw public housing sales and the resulting home ownership as a means for improving housing quality and promoting empowerment among low-income households.

In 1985, HUD launched the Public Housing Homeownership Demonstration Program. According to Silver, "The administration saw privatization of public housing as a way to 'recommodify' housing, reduce federal costs, and improve the local tax base." In addition, the opportunity for home ownership would reduce the dependency of low-income households on the government. From the view of the New Left, however, the self-help overtones of the Demonstration Program reflect "a form of radical, grass-roots democracy that empowers powerless people to resist authority, question the status quo, and . . . challenge the ideology of home ownership in which private property is used for profit."

Based on the experiences of about the first year and a half of the demonstration, Silver concludes that the program "has not significantly privatized this property financially, enhanced self-sufficiency and independence of the buyers, or strengthened their communities." Silver further cautions self-help advocates that home ownership for low-income households requires a public subsidy. As a final critical concern, Silver points out that public housing sales reduce the supply of a precious public resource, which is particularly needed at a time of low-income housing scarcity. As long as there is an acute need for low-rent housing in so many parts of the country, and unless the government is actively promoting the production and rehabilitation of housing, it is difficult to imagine many instances in which the benefits of public housing sales would outweigh the costs.

Overall, there have been many negative aspects of private sector involvement with public housing. Although resident management of public housing is certainly not a negative initiative, it fulfills a function that may be more appropriately filled by the local housing authorities themselves. Thus privatization in this case may divert attention from the more critical issue of how to provide the resources and management capabilities to ensure high-quality services to public housing tenants.

Concerning the sale of public housing, the extreme of privatization, we must be very cautious about the loss of what may be, at least in the short term, an irreplaceable public asset.

It has been observed that public housing presents a paradox: "Nobody likes public housing except the people who live there and those who want to get in" (Rabushka & Weissert, 1977, p. xvi). Homelessness, the lack of affordable housing in many locales, and long public housing waiting lists are key reasons we need a well-managed and well-funded public housing program.

REFERENCES

Bratt, R. G. (1989). *Rebuilding a low-income housing policy.* Philadelphia: Temple University Press.

Friedman, L. (1968). *Government and slum housing.* Chicago: Rand McNally.

Keith, N. S. (1973). *Politics and the housing crisis since 1930.* New York: Universe.

Kolodny, R. (1986). The emergence of self-help as a housing strategy for the urban poor. In R. G. Bratt, C. Hartman, & A. Meyerson (Eds.), *Critical perspectives on housing.* Philadelphia: Temple University Press.

Manpower Demonstration Research Corporation. (1981). *Tenant management: Findings from a three-year experiment in public housing.* Cambridge, MA: Ballinger.

Meehan, E. J. (1979). *The quality of federal policy making: Programmed failure in public housing.* Columbia: University of Missouri Press.

President's Committee on Urban Housing. (1968). *Report of the President's Committee on Urban Housing* (E. F. Kaiser, Chair). Washington, DC: Government Printing Office.

Rabushka, A., & Weissert, W. G. (1977). *Caseworkers or police? How tenants see public housing.* Stanford, CA: Hoover Institution Press.

7

Privatization, Self-Help, and
Public Housing Home Ownership
in the United States

HILARY SILVER

THE PRIVATIZATION OF AMERICAN low-income housing programs in general, and the sale of public housing in particular, long
preceded the Reagan administration (Silver, McDonnell, & Ortiz,
1985). Yet the apparent success of Mrs. Thatcher's "right to buy" policy,
resulting in the sale of more than a million British council houses, gave
the policy new impetus in the United States during the 1980s. Responding to advocates from New Right think tanks and citing the British
experience, the President's Commission on Housing proposed public
housing sales in its 1982 report. Ronald Reagan himself endorsed the
policy in his 1985 State of the Union Address. That year, the Department
of Housing and Urban Development (HUD) initiated the three-year
Public Housing Homeownership Demonstration to assess the conditions under which sales could succeed in an American context. This
chapter evaluates the demonstration's results as of the summer of 1987.

While British researchers soon discovered some of the adverse fiscal
and social consequences of the right to buy (see Forrest & Murie, 1988),
many American members of Congress and housing experts warned that
the tenantry, scale, and quality of British and American public housing
were very different (e.g., Roistacher, 1984). Thus the Democratic-
controlled Congress repeatedly rejected an Urban Homestead Act pro-

AUTHOR'S NOTE: I gratefully acknowledge the support of the A. A. Taubman Center
for Public Policy and American Institutions, Brown University. I also wish to thank Tom
Anton, Bill Gormley, Steve Redburn, and the editors of this volume for their comments.
Special thanks go to Judith McDonnell for her contributions to this research.

posed by Jack Kemp, then a representative from New York active in promoting New Right legislation, which sought to grant a comparable "right to buy" to U.S. public housing residents. However, after considerable debate and amendment, Kemp won support from many liberal Democrats for a limited version of the sales policy included in the 1987 Housing and Community Development Act. It permits successful resident management corporations to purchase their homes after three years. Kemp, who became secretary of HUD under George Bush, has now proposed a HOPE (Homeownership and Opportunity for People Everywhere) program, which offers subsidies for public housing tenants to rehabilitate and convert their homes to resident-owned cooperatives. Yet the evidence that such policies work in the United States has yet to be produced. Early results from the HUD Public Housing Homeownership Demonstration suggest these plans are overly optimistic.

The demonstration initially involved the sale of about 2,000 units in 18 urban and rural sites—16 public housing authorities (PHAs) and 2 Indian housing authorities. HUD continued debt payments on the original construction and provided $500,000 worth of technical assistance to PHAs. But to encourage creative solutions adapted to local conditions, local authorities had discretion over which units to sell, the type of ownership, means of finance, and use of the proceeds. This procedural diversity largely precluded an evaluation research design that could yield generalizations appropriate for extending a sales policy to the entire nation. Indeed, some concluded that "like many experiments, it is rigged to get the desired results" (Dreier, 1986, p. 14).

Yet HUD imposed a number of requirements for all participating PHAs that somewhat limited local discretion. The units sold must be in sound physical condition or already in the process of modernization. Buyers must be sitting tenants or others in public housing who can afford the full costs of home ownership—mortgage, utilities, taxes, insurance, and routine maintenance—under the terms offered. At the same time, involuntary displacement of sitting tenants was prohibited. The PHA also had to specify how sales would affect the long-term availability of low-income housing in the area. However, unlike the 1987 act, the demonstration did not stipulate that replacement units had to be found. Nor were resales restricted to a particular income group, although short-term procedures for preventing windfall profits had to be devised. Indeed, in order to provide incentives for the buyers to make improvements, the policy encouraged profits on resale after at least five years from the date of purchase. Another requirement was the identification of racial and ethnic characteristics of current tenants and the neighborhood to prevent racial discrimination and segregation. Finally, HUD mandated training and counseling for purchasing tenants, as well as the participation of local government and tenant groups.

Despite these safeguards, it appeared by mid-1987 that the avowed aims of the demonstration were not being achieved in virtually any of the sites. In the next section, I discuss the objectives that conservatives and liberals expected public housing sales to accomplish. They included the reduction of government spending, the encouragement of individual self-sufficiency, and the strengthening of communities. The research results that follow suggest that the demonstration, as it is currently structured, does not promote these forms of privatization and self-help. Rather, I conclude that the motivations for promoting public housing sales derive from unconfirmed assumptions—on both the right and the left—associated with the "ideology of homeownership" (Kemeny, 1981; Perin, 1977).

THE OBJECTIVES OF
PUBLIC HOUSING SALES

The Reagan administration's broad domestic priority was to reduce government spending, taxation, and social dependency. Expenditure cutbacks in general fell disproportionately on the working poor (Meyer, 1984), who, it was argued, could achieve self-sufficiency through individual effort. The Reagan administration particularly cut spending on housing programs. New public housing starts fell precipitously, from over 45,000 to only 7,000 units. Rents were raised to 30% of tenant income. While existing subsidies on privately owned low-income rental housing began to expire, tax incentives for new investment of this kind were eliminated (Bratt, 1987; National Low Income Housing Preservation Commission, 1988). Housing assistance increasingly came in the form of housing vouchers to be redeemed in tight private rental markets (Gilderbloom & Appelbaum, 1988, p. 77). More and more PHAs contracted out for services (U.S. Department of Housing and Urban Development, 1983). Although some states and cities initiated their own housing programs to compensate for the loss of federal funds (Stegman & Holden, 1987), others demolished existing public housing or sold it to private developers in order to raise revenues or reduce costs (Bureau of National Affairs, 1986; Connerly, 1986; Hartman, 1986; National Housing Law Foundation, 1986).

The sale of public housing, like these other forms of privatization, aimed to reduce social welfare expenditures and dependency on government. HUD's stated objectives for the Homeownership Demonstration roughly fell into these two categories. By helping low-income families "share in the American goal" of owning their homes and by permitting tenants managing their projects to "take the next step" to ownership, sales would "break the cycle of dependency." They would build "a sense of responsibility," equity to leave to one's children, and a "stake in the community" that would improve neighborhoods and the quality of life even for families remaining as public housing tenants. At the same time, sales would reduce federal costs and control, while enhancing local tax bases (*Federal Register*, 1984, p. 43029; U.S. Department of Housing and Urban Development, 1984).

Some of the conservatives' goals resonated with the values of nonprofit community-based housing groups, including urban homesteaders, sweat-equity participants, low-income cooperatives, and public housing tenant associations. In this view, sales are a variation on the themes of community control and self-help that act to empower low-income populations. In contrast, "public housing is, for the most part, repressive. Typically, the residents have little control over the basic decisions that affect their lives, . . . contributing to a pervasive alienation that is reflected in poor home maintenance, rent delinquency, and even vandalism" (Gilderbloom & Appelbaum, 1988, p. 75). Thus self-help housers' antibureaucratic attitudes and disillusionment with existing housing programs bear a resemblance to the New Right's populist appeals against big government. They feel that housing policy should promote resident autonomy, control, independence, initiative, and self-respect. Indeed, low-income tenants in general are overwhelmingly interested in purchasing homes (Heskin, 1983).

Given this self-help approach, it is perhaps less surprising that some liberal Democrats and housing professionals supported, with reservations, an essentially Republican policy of public housing sales. However, the analogy with neoconservatism should not be taken too far. The New Left interprets the self-help housing movement's aspiration for control as community, rather than individually, motivated (Peterman, 1988). Rather than promoting home ownership and private property for reasons of personal profit, it sees them as ways to improve housing for those actually using it and to resist the authority of landlords. Housing activists may form networks with other low-income organizations to pool resources in the quest for neighborhood autonomy (Borgos, 1986; Lawson, 1986).

If the New Right and the New Left tend to agree that home ownership will help public housing tenants become more self-sufficient, the New

Right also expects it to save the government money. This is why some observers have expressed grave political misgivings about lionizing self-help efforts. They worry that self-help diverts political attention from the massive scope of the low-income housing problem and may play into the hands of fiscal conservatives who wish to unload government's obligations onto the poor (Katz & Mayer, 1985; Schuman, 1986). The provision of limited subsidies may co-opt self-help groups or divide them as they compete for resources (Lawson, 1986).

Indeed, empirical evidence indicates that self-help housing works only under special circumstances. Its most impressive successes result from efforts developed not by outsiders from the top down, but from the bottom up as a practical last resort in response to severe housing crises (Harloe, 1978; Kolodny, 1986). Evaluations of tenant management indicate that it is, at best, of mixed success (Hundley, 1985; Peterman, 1988). These projects fail in the long run without strong, continuous leadership, long periods of training, external technical assistance, and task-oriented responsibilities appealing to self-interest or community sentiment (Center for Urban Policy Programs, 1975; Diaz, 1979; Rigby, 1985; U.S. Department of Housing and Urban Development, 1979).

Opinions differ on whether self-help is more expensive than conventional operations (Kolodny, 1986). Public housing resident management groups claim that they improve rent collections, reduce vacancies, and lower administrative costs (National Center for Neighborhood Enterprise, 1984; Rigby, 1985). But evaluations of a demonstration program in the late 1970s concluded that tenant management corporations (TMCs) cost more than conventional PHAs (Diaz, 1979; U.S. Department of Housing and Urban Development, 1979). Most self-help projects of the very poor succeed only with public subsidies, tax forgiveness, and technical assistance (Borgos, 1986; Dreier, 1986; Henig, 1985, pp. 96-200; Lawson, 1986; Schuman, 1986; Varady, 1986). Finally, making existing properties available to poorer self-help groups does little to alleviate the great scarcity of low-income housing. Nonfederal housing programs produce "only a small fraction of the body of the federal programs that this activity is trying to replace" (Stegman & Holden, 1987, pp. 4-5).

These concerns may explain why only a few dozen of the more than 3,000 PHAs even applied for the demonstration. Moreover, those participating pursued a variety of goals that, in some cases, varied from those of the administration. Interviews conducted for this study revealed that while one-third the officials overseeing the demonstration endorsed public housing sales for reasons similar to HUD's, others simply saw the program as an opportunity to address their own local administrative or managerial concerns. For example, sales offered PHAs flexibility with the composition of their stock, some of which may be underused or difficult and costly to manage. Indeed, almost half the PHAs participating in the demonstration had some reservations about it. They qualified their endorsement with concerns about replacement units, characteristics of the purchasers, or the type of housing suitable for sale.

In light of these reservations, it is instructive to evaluate the Public Housing Homeownership Demonstration in terms of the goals it sought to achieve. Both conservatives and some liberals expect sales will increase tenant self-sufficiency and strengthen communities. The New Right also sees in them a basis for a nationwide housing policy that saves government money and helps "privatize" public housing. The research described below examines whether these claims are justified.

A PRELIMINARY APPRAISAL
OF THE DEMONSTRATION

METHODS

Most of the data for what follows come from one-hour interviews with PHA officials overseeing the HUD demonstration in participating localities. After pretests, personalized advance letters and questionnaires asking for general data on public housing in the locality were sent by mail. The actual interviews took place in June 1987, with several follow-ups in the fall. By this time, 165 units had been sold nationwide in nine cities, with an additional 42 units scheduled for sale before the first of the year. In addition, 4 PHAs (including the 2 Indian authorities) had dropped out of the program and 2 new ones were approved. The data thus refer to 16 participating PHAs.

Supplementing the interview data is information from PHA applications to participate in the demonstration, congressional testimony, and government documents. I also conducted a number of in-depth interviews with public housing tenants purchasing or managing their homes in Chicago and Washington, D.C.

PUBLIC HOUSING SALES AND
PRIVATIZATION: FISCAL IMPACTS

While conservatives view public housing sales as a form of privatization, a case can be made that the demonstration is actually *reprivatizing* housing. In most cities, the units for sale are scattered over a number of sites. Frequently, PHAs originally acquired these properties from the private sector, although they were more difficult and costly to operate and maintain than single-site projects. Ironically, these units are now being reconverted to their original use as single-family private homes, albeit with owners of a different social class. Other PHAs are selling units they built and failed to sell under Turnkey III, a low-income home ownership program initiated in the 1960s. Given HUD's selection of sites, it is difficult to test whether high-rise public housing is amenable to resident home ownership.

With respect to public *finance*, there are stronger reasons to doubt that the demonstration privatizes public housing (see Table 7.1). First, federal debt service on the buildings' initial construction will continue after sale (Schussheim, 1984, p. 17). Second, actual mortgage lending practices indicate that economic dependence on government will persist even after "sales" are transacted. Where sales have yet to proceed, PHAs are still seeking or have obtained commitments from private banks. But many localities reported difficulty in securing private loans and were forced to revert, at least for the time being, to some form of public finance. In sites where sales have already occurred, state housing agencies, PHAs, the FHA, or below-market lending institutions provided the mortgages. Even apparent exceptions to this rule entailed some public involvement in the loans, whether by tapping into low-interest earmarked funds, indemnifying the risk, or actually holding title to a private mortgage for the "buyers." Two PHAs expected to sell through a lease-purchase arrangement that only gradually reduces public obligations. Finally, most PHAs have extended a silent second mortgage on top of the primary one. Besides temporarily preventing windfall profits on resales, it reduces down payments and is forgiven in the long run, thus subsidizing the purchase price. In sum, sales have proceeded more quickly where the *public* sector continues to subsidize *privatization*.

Third, conservative notions of privatization imply that the sales should save the government money. While a rigorous fiscal analysis of the demonstration is impossible due to the great diversity among sites, the data presented in Table 7.1, assessing at least some sources of outlay and revenue, indicate that public sector costs may outweigh the savings. While a number of assumptions were made in making these estimates,[1] I approximate that government will save about $4.6 million a year through reduced operating costs, interest payments, and higher local taxes. But it will lose at least $12.1 million a year in tax expenditures,

TABLE 7.1

Government Costs and Savings from the Sale of U.S. Public Housing

Revenues	*Expenditures*
One-time Savings	One-time/short-term costs
Proceeds (none in most sites, but earmarked for residents anyway)	local costs incidental to sales (no down payment, closing costs, legal fees, training, community development $, tenant relocation, staff time)
	HUD administration: $2.9 million
	modernization: $15.5 million
	tax abatements (3 sites)
Continual savings (annual)	Continual losses (annual)
local property taxes: $948,000	PILOTs
operating costs $3,300,000	tax deductions: $200,000
interest on public mortgages in 11 sites, roughly equal to rents early on, then declines generous estimate: $393,000	rents on non-publicly financed units: $3.4 million
	debt payments: $8.5 million
	replacement units/vouchers
Total	Total
about $4.6 million a year	at least $18.4 million one-time costs and about $12.1 million a year

rents, continued debt payments, and administrative expenses. If one-time costs are added, the outlays could rise another $18.4 million, with only the sales proceeds to balance them. However, unlike in Britain, where proceeds gave the Exchequer a short-term windfall (see Forrest & Murie, 1988; Kilroy, 1982), U.S. sales proceeds are being earmarked either for rehabilitating the units or, more typically, as buyer reserve funds.

Even HUD (1984) has acknowledged that, if expanded to a national program, public housing sales could cost the federal government more than they save, depending on the conditions of sale. If subsidies were increased to raise the number of sales, anything beyond a nominal program would be more expensive than retaining the units for rent.[2]

To conclude, the sales demonstration does to a large extent privatize the maintenance of public housing, but it does not privatize its financing. Given the reluctance of private sector lenders to assist low-income buyers, this "privatization" program falls short of the goal of reducing public intervention in the housing market.

In one respect, however, the fiscal objective may have been met. Given that commitments for construction debt would be ongoing with or without sales, the administration may indeed be saving *federal* money. The one-time expense of the demonstration and the relatively low value of tax deductions claimed by low-income families are more than offset by reduced operating subsidies. The public costs of the demonstration fall primarily on *PHAs*. They do not directly benefit from local property taxes, but where they are not the primary mortgage lenders, PHAs do lose rents from relatively affluent tenants and modernization funds that could be used to improve the remaining housing stock. As in other "new federalism" proposals of the Reagan administration, federal costs are imposed on localities. This may also explain some of the reservations expressed by PHAs about the demonstration.

PUBLIC HOUSING SALES
AND SELF-HELP

Observers of self-help housing have long recognized that public subsidies are necessary to make such programs work. Nevertheless, these expenditures might be justified as a means of enhancing residents' control over their homes and strengthening low-income communities. The Public Housing Homeownership Demonstration also incorporated these self-help goals.

For example, the requirement that tenants participate in the sales program might appear to offer an avenue of empowerment. However, in reality, tenants had very little involvement in its design or execution. In part, this is because there were no existing tenant associations among buyers in the majority of sites. Rather, the PHAs assembled buyers they deemed eligible for home ownership.

Indeed, three-fourths of the cities are selling single-family houses, and, in at least nine cases, the houses are scattered among many sites. Multifamily housing will be sold to tenant cooperatives or condominiums in only five cities. However, these units are mostly in low-rises, without the critical mass helpful in organizing associations in buildings with greater population density. Moreover, the conditions usually conducive to tenant-initiated mobilization were absent. Far from deteriorating to the point of crisis, most of the housing for sale was recently modernized and, in one case, built from scratch. By mid-1987, the cooperatives, with one exception (see Sarlo, 1987), were still being organized and trained. As in other self-help programs, these have been difficult to organize from the top down.

Mandated training classes provided the bulk of any tenant "participation" in the demonstration. The training required for cooperative or condominium ownership is extensive, particularly in areas of the country where such tenures are unfamiliar. But virtually all tenants received training, even in Nashville and Washington, where purchasing residents were originally selected for the Turnkey III program and thus were oriented to the demands of ownership.

It is perhaps surprising that no tenant management corporations were selected to participate in the HUD Public Housing Homeownership Demonstration, given that one of its initial goals was "providing tenants who are already involved in the management of their public housing projects the opportunity to take the next step to ownership." Indeed, conservative Republicans advocating public housing sales frequently cited TMCs as proof that "local initiatives" work. Being excluded from the demonstration, these resident groups lobbied Congress for approval and assistance to buy their buildings. The 1987 Housing and Community Development Act granted this, but required a three-year waiting period before any sales could be transacted.

Thus the sales demonstration itself did not significantly empower existing public housing resident communities. On the contrary, one could argue that it even disrupted them. For one thing, most buyers had incomes over twice the public housing average. AFDC recipients, who constitute about half the national public housing population (Council of Large Public Housing Authorities, 1986), were frequently disqualified from buying their homes, even if their incomes were above the eligibility threshold. In this way, the demonstration deprived the "undeserving poor" who remained tenants the same benefits received by their "working poor" neighbors. The restriction may have been deliberately designed to provide work incentives for dependent families who really want to buy their homes, but in at least one sales site, it led to some resentment on the part of welfare recipients.

The sales disrupted communities in yet another way. While the demonstration explicitly prohibits involuntary displacement of nonbuying tenants, some PHAs exerted considerable pressure on them to move. This is because PHAs first identified the housing to be sold and only later the tenants who could buy it. In a number of places, this resulted in relocation and turnover in the community. In Denver, where modernization and demolition preceded the sales, all residents were initially relocated, but renovation left fewer units than before. When only the buyers were allowed to move back in, nonbuying tenants who wanted to return to their community protested. Elsewhere, nonbuyers complained to lawyers of PHA harassment to relocate. Tenants living in units designated for sale and in violation of public housing rules quickly came to PHA attention for eviction. This seems a far cry from "improving the quality of life for both those families remaining as tenants of public housing and those who move into homeownership" (*Federal Register*, 1984).

Despite these instances, most PHAs took a tolerant approach toward tenants who chose not to relocate, indicating they "do not want to push them." Others work with nonbuying tenants to help them eventually qualify for purchase. In condominium and cooperative settings, future research may indicate whether buyers help each other move to co-ownership. However, groups organized from the outside may just as soon develop conflicts as harmony.

The demonstration also sought to build "a sense of responsibility and a homeowner stake in the community that would lead to neighborhood stability and ultimate improvement." Yet there are reasons to doubt that community upgrading will spill over from public housing sales. Where scattered-site housing is sold, home ownership should have a negligible effect in the neighborhood. Moreover, evaluations of similar low-income home ownership programs, such as Urban Homesteading, have found little or no impact on neighborhood revitalization (Varady, 1986).

This does not mean, however, that areas *already* under pressure from rising property values will not benefit from public housing sales. In fact, in the majority of neighborhoods in which public housing is being sold, property values have been rising; they are declining in only a few. Although most of the sites currently are in predominantly Black areas, many are close to downtown, historical districts or, in the case of the Virgin Islands, the waterfront. Thus there is a certain irony in arguing that public housing sales will lead to neighborhood improvement when the properties are already appreciating. Indeed, the President's Commission on Housing argued that sales are a good way to end government ownership in revitalizing neighborhoods. Elsewhere in the country, PHAs have already sold entire projects in gentrifying areas to recoup their enhanced equity (Hartman, 1986). Thus, rather than promoting revitalization, public housing sales may be a good method of preventing displacement of low-income minority families from areas already in the process of gentrification.

Since property values are rising, it was fortuitous that the demonstration required a windfall profits provision that prevented resale of units for at least five years. Some PHAs went further, extending the limitation for seven or ten years or even for the life of the mortgage. In the vast majority of cities, windfall profits were restricted through either a "silent second mortgage" due on resale or the PHA's first right to refuse purchase. Three of the five cooperatives have "limited-equity" provisions restricting all resales to low-income households.

Yet windfall profit limitations appear to conflict with the conservatives' goal of encouraging social mobility. Indeed, low-income home ownership may hinder it. The five-year or more profit moratorium may restrict the geographical mobility of purchasing families. Since most qualified buyers already held jobs, albeit low-wage insecure ones, home ownership alone was unlikely to serve as a work incentive. In fact, the new responsibilities of managing a home may demand time, effort, and resources that could otherwise be spent at work. To be sure, buyers may now control decisions about their homes previously regulated by the PHA, but "doing it yourself" requires discretionary income that many families simply do not have (Pahl, 1984). So far, however, buyers have continued to rely on the PHA for repairs or at least for advice on how to do it themselves.

Supporters of public housing sales assumed that home ownership provides buyers with some capital to bequeath to their children. Building this "home owner stake" would seem to entail some investment and potential profit. Yet the required down payments were negligible, and equity was designed to build up so slowly that buyers may prefer to walk away rather than lay out any savings on emergency expenses. Ironically, rather than worrying about equity to leave to their children, some tenants expressed concern that, under current PHA rules about "doubling up," they would not be able to *house* their adult children securely unless they bought. Indeed, the demonstration seemed more concerned with providing capital gains in the long run than a safety net in case of defaults in the short run.

It is important to recall that most public housing home buyers hold relatively low-wage, insecure jobs with meager benefits in the event of unemployment or disability. Many live in multigenerational households with unemployed children or parents. By purchasing their homes, they must now pay slightly more of their incomes on housing than they did as tenants, stretching their family budgets. Thus extraordinary costs may at any time lead to mortgage or tax defaults. Indeed, previous low-income home ownership programs, such as Section 235, suffered from defaults incurred when unexpected maintenance problems or rising utility costs strained buyers' financial resources (Schafer & Field,

1980). Similarly, among New York's relatively successful low-income cooperatives, about half of the older ones fell far behind in their debt service and tax payments (Lawson, 1986, p. 267). In many of the demonstration sites, proceeds will be placed in escrow as a loan fund for future catastrophic costs. However, when PHAs were questioned about procedures in the eventuality of defaults, most had no plans to deal with the situation. Some responded that they did not want to think negatively. Others expected that they or the banks would repossess the property, and the dislocated home owners and their dependents could rejoin the long public housing waiting lists. Thus buying public housing on an insecure income may leave families worse off than when they started. More than home ownership, household financial security contributes to low-income community stability and individual autonomy. Indeed, many successful self-help housing projects also provide employment and job training, an approach absent from the home ownership demonstration.

In sum, the sales demonstration is unlikely to confirm the expectation of self-help housers that it can significantly empower low-income individuals and existing communities. Indeed, the program's goals appear to assume that public housing tenants have no stable neighborly relations, but that home ownership, by bestowing individual economic stakes, would necessarily create them. Existing communities and neighborly relations were subordinated to new ones that PHAs would build among selected buyers.

CONCLUSIONS

The major conclusion of this early evaluation of the HUD Public Housing Homeownership Demonstration is that it is unlikely to achieve the objectives that conservative Republicans and some liberal Democrats expected of it. The sale of public housing, at least as the demonstration structured it, has not significantly privatized this property financially, enhanced the independence of the buyers, or strengthened their communities. Rather, these goals reflect the pervasive ideology of home ownership, which in turn buttresses federal housing policies that favor it over public renting.

Just as the British right-to-buy program resulted in long-term government fiscal losses (Kilroy, 1982), the American conservatives' belief that privatization would reduce government spending and buyers' dependence on it found little support. Any savings were more than offset by new fiscal burdens on PHAs. The financial costs of the demonstration appear even less justifiable in light of further issues, raised in

British assessments of the right to buy, concerning the effect of sales on those remaining in the public sector (e.g., Forrest & Murie, 1988). First, there are dubious equity implications of subsidizing sales to relatively more privileged public housing tenants and "creaming off" residents whose rents might otherwise cross-subsidize the stock. Second, British critics have noted the implications of sales for both the physical and the socioeconomic profile of public housing. Since the American program requires that sound or rehabilitated housing be sold, PHAs lose some of their best stock as well as modernization funds that could have repaired units remaining in the public sector. While the average income of public housing tenants keeps falling, sales further distinguish the working from the dependent poor. Moreover, the demonstration maintains current racial segregation patterns by doing nothing to encourage these predominantly minority buyers to obtain housing in White areas.

Finally, while the scale of the U.S. program pales beside the British one, the existing supply of public housing is particularly precious in a period of burgeoning demand for it. The million British right-to-buy sales shrank the council sector to 6 million units, or from a third to 26% of the national housing stock. In contrast, the HUD demonstration involved about 1,300 public housing units, or a minuscule proportion of the 1.3 million total. Even prior sales programs, which amounted to a loss of fewer than 4,000 units (Lundquist, 1984), had little effect on the overall supply. However, subsidized housing is scarce in the United States, representing at most 3% of all housing units, at a time when homelessness, private rent burdens, and public sector waiting lists are rising, even in the sites participating in the sales program (Apgar & Brown, 1988; Clay, 1987). Any nationwide sales policy should consider the effects on supply. For example, the 1987 act required one-for-one replacements for any property sold.

Given that the demonstration is unlikely to achieve its aims and has additional drawbacks for remaining public housing tenants, one must ask why the policy of public housing sales has been pressed onward, even before the evidence could be evaluated. One reason may be that the stated goals of the demonstration were never seriously intended to serve as a basis for assessment. Rather, they simply reflect ideological assumptions about home ownership made by conservatives and liberals alike.

The myths of home ownership include "the psychological and 'natural' desire of owning, its inherent security of tenure (and by implication the inherent insecurity of other forms of tenure), and the capital asset which is produced" (Kemeny, 1981). Just like stages of the life cycle, the progression from renting to owning is construed as natural, so that anyone, regardless of income, should "take the next step" to attain it. People who remain tenants for too long deviate from the "correct chronology of life." Perin (1977), citing a former secretary of HUD, illustrates the ideology of home ownership in terms very reminiscent of the demonstration's self-help goals:

> The family who owns its own home, not only has an investment in a house, it has an incentive to take an active role in the decisions which shape its neighborhood. . . . It is axiomatic that when neighborhoods turn from "owner" to "rental" properties, evidence of neglect begins to show almost immediately. The reverse is also true. . . . Homeownership provides a sense of identity, of roots and of security, which . . . protects against social alienation. (p. 78)

Or, in the current secretary's words:

> Public housing discourages work and saving, raises numerous barriers to the upward mobility of tenants, and sometimes has degenerated into dilapidated and depressing slums. . . . The mere act of homeownership transforms tenants, giving them a new sense of belonging and self-reliance. (Kemp, 1984)

This conception of home ownership, in which community flows directly from private property interests and is therefore lacking among public housing tenants, has shaped American housing policies and is in turn reinforced by it through differential subsidies (Kemeny, 1981; Perin, 1977). Thus the fiscal cost of the sales and their detrimental effects on the remaining public tenantry are actually consistent with the government's regressive subsidization of American home owners.

The notion of community rooted in home ownership stakes is not the same as that envisaged by most self-help housers. Nevertheless, their liberal Democratic supporters, in the belief that owner-occupiers have greater autonomy, control, and security than tenants, have lent support to a conservative Republican policy that has provided a rationale for substantial low-income housing cutbacks. Local tenant activities, undertaken as a last resort in the face of government inaction to provide affordable housing, are being confused with a progressive new approach to national housing policy. Any such approach should be mindful of the evidence, presented in this chapter, that low-income home ownership requires substantial government subsidies and technical support. For poorer Americans, neither privatization nor self-help is absolute.

NOTES

1. Details of the cost analysis, including financial assumptions made, are found in McDonnell and Silver (1987), available from the author on request.

2. A similar analysis of Kemp's right-to-buy bill showed similar fiscal consequences (Schussheim, 1984).

REFERENCES

Apgar, W., & Brown, H. J. (1988). *The state of the nation's housing, 1988.* Cambridge, MA: Joint Center for Housing Studies.

Borgos, S. (1986). Low-income homeownership and the ACORN squatters campaign. In R. G. Bratt, C. Hartman, & A. Meyerson (Eds.), *Critical perspectives on housing* (pp. 428-446). Philadelphia: Temple University Press.

Bratt, R. G. (1987). Private owners of subsidized housing vs. public goals. *Journal of the American Planning Association, 53*(3), 328-336.

Bureau of National Affairs. (1986). *Housing and development reporter.* Washington, DC: Author.

Center for Urban Policy Programs. (1975). *Tenant management corporations in St. Louis public housing: The status after two years.* St. Louis: St. Louis University.

Clay, P. L. (1987). *At risk of loss: The endangered future of low-income rental housing resources.* Washington, DC: Neighborhood Reinvestment Corporation.

Council of Large Public Housing Authorities. (1986). *Public housing today.* Washington, DC: Author.

Connerly, C. (1986). What should be done with the public housing program. *Journal of the American Planning Association, 52,* 142-155.

Diaz, W. A. (1979). *Tenant management: An historical and analytical overview.* Washington, DC: Department of Housing and Urban Development, Manpower Demonstration Research Corporation.

Dreier, P. (1986, August 4). Private project: Public housing for sale. *New Republic, 195,* 13-15.

Federal Register. (1984, October 25). Department of Housing and Urban Development Public Housing Homeownership Demonstration. Vol. 49, 208, 43028-34.

Forrest, R., & Murie, A. (1988). *Selling the welfare state: The privatization of public housing.* London: Routledge.

Gilderbloom, J. I., & Appelbaum, R. P. (1988). *Rethinking rental housing.* Philadelphia: Temple University Press.

Harloe, M. (1978). *Housing management and new forms of tenure in the United States.* London: Center for Environmental Studies.

Hartman, C. (1986). Housing policies under the Reagan administration. In R. G. Bratt, C. Hartman, & A. Meyerson (Eds.), *Critical perspectives on housing* (pp. 362-377). Philadelphia: Temple University Press.

Henig, J. (1985). *Public policy and federalism.* New York: St. Martin's.

Heskin, A. D. (1983). *Tenants and the American dream: Ideology and the tenant movement.* New York: Praeger.

Hundley, R. (1985, May). *The promise of resident management: A step toward homeownership*. Paper presented to the OECD Seminar on Community Involvement in Urban Service Provision, Paris.

Katz, S., & Mayer, M. (1985). Gimme shelter: Self-help housing struggles within and versus the state in New York City and West Berlin. *International Journal of Urban and Regional Research, 9*(1), 15-46.

Kemeny, J. (1981). *The myth of home ownership*. London: Routledge & Kegan Paul.

Kemp, J. (1984, September 27). *Testimony before the Joint Economic Committee, U.S. House of Representatives*. Washington, DC: Government Printing Office.

Kilroy, B. (1982). The financial and economic implications of council house sales. In J. English (Ed.), *The future of council housing*. London: Croom Helm.

Kolodny, R. (1986). The emergence of self-help housing as a housing strategy for the urban poor. In R. G. Bratt, C. Hartman, & A. Meyerson (Eds.), *Critical perspectives on housing* (pp. 447-462). Philadelphia: Temple University Press.

Lawson, R., with Johnson, R. (1986). Tenant responses to the urban housing crisis, 1970-1984. In R. Lawson with M. Naison (Ed.), *The tenant movement in New York City, 1904-1984* (pp. 209-271). New Brunswick, NJ: Rutgers University Press.

Lundquist, W. (1984, August 3). *Survey of Section 5(h) and Turnkey III Sales of Public Housing to Tenants*. Washington, DC: U.S. Department of Housing and Urban Development.

McDonnell, J., & H. Silver. (1987). *The HUD Public Housing Homeownership Demonstration: A preliminary appraisal*. Paper presented at the meetings of the Association of Collegiate Schools of Planning, Los Angeles.

Meyer, J. A. (1984). Budget cuts in the Reagan administration: A question of fairness. In D. L. Bawden (Ed.), *The social contract revisited* (pp. 3-64). Washington, DC: Urban Institute Press.

National Center for Neighborhood Enterprise. (1984). *The grass is greener in public housing: From tenant to resident to homeowner—A report on resident management in public housing*. Washington, DC: Author.

National Housing Law Foundation. (1986, May/June). HUD rejects proposal to sell public housing. *Housing Law Bulletin,* pp. 14-15.

National Low Income Housing Preservation Commission. (1988). *Preventing the disappearance of low income housing*. Washington, DC: National Corporation for Housing Partnerships.

Pahl, R. (1984). *Divisions of labour.* Oxford: Basil Blackwell.

Perin, C. (1977). *Everything in its place: Social order and land use in America*. Princeton, NJ: Princeton University Press.

Peterman, W. A. (1988, May/June). Resident management: Putting it in perspective. *Journal of Housing,* pp. 111-115.

Rigby, R. (1985). A community based approach to salvaging troubled public housing: Tenant management. In N. Prak & H. Priemus (Eds.), *Post-war public housing in trouble* (pp. 19-34). Delft: Delft University Press.

Roistacher, E. (1984, Autumn). A tale of two conservatives: Housing policy under Reagan and Thatcher. *Journal of the American Planning Association,* pp. 485-492.

Sarlo, S. (1987, November/December). Focus on Denver: Award-winning PHA Program. *Journal of Housing,* pp. 225-231.

Schafer, R., & Field, C. G. (1980). Section 235 of the National Housing Act: Homeownership for low-income families? In J. Pynoos, R. Schafer, & C. Hartman (Eds.), *Housing urban America* (2nd ed., pp. 485-496). New York: Aldine.

Schuman, T. (1986). The agony and the equity: A critique of self-help housing. In R. G. Bratt, C. Hartman, & A. Meyerson (Eds.), *Critical perspectives on housing* (pp. 463-471). Philadelphia: Temple University Press.

Schussheim, M. J. (1984, December 14). *Selling public housing to tenants: How feasible?* Washington, DC: Congressional Research Service, Library of Congress.

Silver, H., McDonnell, J., & Ortiz, R. (1985, November/December). Selling public housing: The methods and motivations. *Journal of Housing,* pp. 213-228.

Stegman, M., & Holden, J. D. (1987). *Nonfederal housing programs: How states and localities are responding to federal cutbacks in low-income housing.* Washington, DC: Urban Land Institute.

U.S. Department of Housing and Urban Development. (1979). *The National Tenant Management Demonstration: Status report through 1978.* Washington, DC: Office of Policy Development and Research, Housing Management and Special Users Group.

U.S. Department of Housing and Urban Development. (1983, May). *Public housing authority experience with private management: A comparative study.* Washington, DC: Government Printing Office.

U.S. Department of Housing and Urban Development. (1984, April 6). *Issue paper on the sale of public housing to tenants.* Washington, DC: Government Printing Office.

Varady, D. (1986). *Neighborhood upgrading: A realistic assessment.* Albany: State University of New York Press.

8

The Privatization of
Public Housing in Israel

ELIA WERCZBERGER

THE PROBLEM

Privatization in the general sense means a reduction of the role of the state. It may entail the shift from public intervention to private action along any of three dimensions of government activity: provision, subsidy, and regulation (Le Grand & Robinson, 1984). Especially in the housing market, there is considerable leeway for selecting the desirable mix of private and public involvement. Privatization policies may thus be concerned with the sale of public housing or only with the modification of certain aspects of the supply of housing services.

The increasing recent emphasis on privatization puts a premium on reducing the role of government and on increased reliance on citizen action in solving societal or individual problems (Glazer, 1983; Silver, McDonald, & Vitiz, 1985). Objectives pursued in the context of the privatization of public housing include, in particular, equity accumulation by the poor and sharing in what serves as a major status symbol. Moreover, it is used as an instrument for the socialization of the tenants to values such as maintenance, thrift, and personal initiative. Underlying most of these efforts is the assumption of the universal desirability

AUTHOR'S NOTE: The research on which this chapter is based was commissioned by the Amidar housing corporation. Its interest and support are gratefully acknowledged. I would especially like to express my gratitude to Dr. B. Rippa, who initiated the study and contributed his experience and knowledge of the TIR program, and to the many community organizers who carried out the interviews. Special thanks are due to Illan Ben-Ami for his general assistance and the administration of the fieldwork. Last, but not least, I would like to mention the editors of this volume, who contributed through criticism and advice to the conciseness and focus of this report.

of home ownership for autonomy and ontological security (Saunders, 1988).

Israeli housing policy has concentrated on two approaches to privatization: the sale of dwellings to the tenants and the shift of the responsibility for maintenance of dwellings to the residents. Today more than half of the former public housing units have been sold to tenants. Privatization of maintenance, on the other hand, has been less successful, as most of the renters have been reluctant or unable to take care of their housing.

This chapter evaluates a renovation program undertaken by Amidar, the largest public housing corporation in Israel, to reverse this trend. The program requires the active participation of the residents in order to promote the privatization of ongoing maintenance and ownership. The partial privatization of one aspect of the provision of housing services is thus justified by its presumed effect on privatization of other aspects of public housing.

RESIDENTS' PARTICIPATION
IN HOUSING PROGRAMS

Self-help housing and resident participation have become quite popular as policies for improving housing conditions of the poor (Burns, 1983; Ward, 1982). Many of these programs are expected to advance additional objectives as well, although their effectiveness is rarely examined. Rosener (1978) suggests that the effectuality of participation depends on two factors. First, there must be a direct cause-effect relationship between participation and objectives. The empirical evidence for such an effect is rather slight. Burns (1983), for example, has observed greater satisfaction with housing and life and less alienation among the residents in self-help housing projects. Expansion of dwellings in public housing by the residents was found to enhance their inclination to make further improvements (Oxman & Carmon, 1986) and to have a positive effect on their attitudes with regard to the neighborhood (Carmon & Gavrieli, 1987). Finally, sharing in renovation cost has been shown to be associated with better maintenance of the shared areas (Werczberger & Ginsberg, 1987).

Second, according to Rosener, there must be agreement on the goals between the residents and the agency. This seems to be difficult to achieve. Planners and politicians typically are motivated by objectives, such as the democratic value of participation, the solution of agency problems, and changes in client behavior or attitudes (Katan, 1974). Participating residents, on the other hand, can be expected to be prompted by personal objectives, such as increased control over their environment, status enhancement, and, not least, the subsidies and assistance offered for participation. Hence contradictions between the objectives of the residents and those of the sponsors are almost inevitable. Unless incentives are provided that make the behavior desired by the agency also attractive for the residents, participation in housing programs is unlikely to achieve its behavioral objectives.

In the following section a brief survey is offered of the development of public housing in Israel, and an examination of the renovation program is presented. The next two sections describe the methodology and the main research findings of this study. The final section evaluates the results in terms of the effectiveness of the program.

PUBLIC HOUSING IN ISRAEL

BACKGROUND AND POLICY

The construction of public housing in Israel began as a response to the catastrophic housing shortage caused by mass immigration between 1948 and 1953.[1] Since then, over 300,000 units have been built, mostly for new immigrants, but also for other groups considered to be deserving. They are administered by several public housing corporations, of which the government-owned Amidar is the largest.

Rents are set based on social and political criteria rather than according to cost or market value. Because of the past reluctance to link rents to the rate of inflation, payments by tenants have declined over the years to a fraction of the market rent. Rent revenues have thus never been sufficient to pay even for administration, let alone for maintenance, of the buildings. As a consequence, the upkeep of the dwellings and the shared areas was left to the residents, except for major repairs, and was thus in effect privatized. Since most of the tenants lacked the resources and motivation requisite for maintenance, the inevitable effect was widespread neglect and blight.

Because of the need for immediate solution and acute economic difficulties, the quality and size of the units built during the early years was even then far from adequate. Thus most of the housing stock constructed in the 1950s and early 1960s is now considered substandard. The combination of low initial quality and lack of maintenance resulted in physical deterioration and the flight of the economically stronger households, accelerating the residualization of public housing. Many of the projects have thus become slums, with all the accompanying symptoms of social pathology and physical decay.

During recent years, an increasing effort has been made to sell the dwellings to the sitting tenants at subsidized prices. Privatization of ownership is expected to enhance motivation for maintenance and to strengthen the stake of the residents in their locality, particularly in development towns. Last but not least, revenues from the sales could be used to avoid budget constraints imposed by the treasury. Nevertheless, the rate of sales seems to be declining, as ownership is less attractive for the remaining tenants, many of whom pay, because of their economic circumstances, only symbolic rents.

In 1986, Amidar owned only about 110,000 dwelling units out of the 292,000 units it had built over the years (Sagiv, 1986). Public housing's share in the housing stock thus decreased from a peak of 23.4% in 1966 to about 10% in 1989, while 75% of all households live now in owner-occupied units (Werczberger, 1988). Moreover, most of public housing consequently consists today of a mixture of publicly and privately owned units.

MAINTENANCE OF PUBLIC HOUSING

The maintenance of the shared areas in public housing and condominiums in Israel is the responsibility of the residents. It is administered by elected and unpaid "building committees," whose main task is the collection of the dues required for the operation, maintenance, and cleaning of shared facilities, such as central heating and stairways. The actual work is usually carried out by small contractors or occasionally by the residents. Management of maintenance tends to be a constant source of tension and frustration, particularly in low-income buildings, because of indifference, mistrust, and lack of effective sanctions against recalcitrant tenants. The situation is worse in public housing, in which efforts to improve maintenance of the shared areas are paralyzed by poverty and confusion about the responsibility for upkeep.

THE PROGRAM FOR
TENANTS' INVOLVEMENT IN RENOVATION

The Program for Tenants' Involvement in Renovation (TIR) was developed for buildings in which at least half of the tenants still rent their dwellings and in which maintenance problems are, as a consequence, particularly severe. The program had five explicitly declared objectives:

(1) improved attitudes of the residents toward the buildings and their maintenance

(2) organization of building committees to be responsible for managing the shared property

(3) improved maintenance of the shared areas in and around the buildings

(4) reduction in renovation costs incurred by the public housing corporation (not investigated in this study)

(5) promotion of the sale of the dwellings to their tenants

The program itself was based on three elements, each of which was considered essential to its success. (a) Paid nonprofessional "building committee organizers" were employed to convince the residents in selected buildings to participate in the program and to organize building committees. They also provided advice during the renovation and functioned as intermediaries between the tenants and the public housing corporation. (b) The constitution of a building committee was a precondition for participation in the program. The tasks of the building committee included making a decision on a renovation program, collection of dues, negotiation with the contractor, and the like. (c) Financing of the renovation was shared equally between the residents and the housing corporation, the approval of which had to be obtained for the renovation program.

The most important provision of the TIR program was the insistence on the direct involvement of the residents in the renovation process through the building committee and cost sharing. The assumption was that active participation would further social integration of the residents and their identification with the building. This was expected to enhance their self-esteem and to reinforce their feeling of being in control and having a stake in their environment. It was thought that the resulting change in outlook would lead to greater demand by tenants for maintenance and ownership. The desirability of owner occupancy for the low-income population living in public housing was, however, never questioned. The purpose of the study reported here was to evaluate the effectiveness of the TIR program in achieving these goals.

RESEARCH METHODOLOGY

Two research instruments were developed for this study: a field survey and a household survey. The field survey evaluated the cleanliness and physical condition of the shared areas in the 192 buildings included in the sample. It yielded an index of the quality of ongoing maintenance, in contradistinction to the condition of the shared area, which is due to renovation. The index is based on five aspects of the semipublic property: lobby, staircase, mailboxes, courtyard, and gardening. The household survey comprised 502 residents living in these buildings. The questionnaire dealt with their personal characteristics and their attitudes regarding their buildings, their neighbors, and maintenance.

Because of the need for feedback during program implementation, a cross-sectional design with control groups was used. Data were collected from ten building clusters, each of which comprised ten TIR buildings and two buildings for each of three control groups. These comprise, respectively, (a) buildings in which organizers had been active, but whose residents, nevertheless, did not join the TIR program; (b) buildings renovated without resident involvement; and (c) buildings in which no agency-sponsored renovation or community work had been undertaken. The respondents in each building were analyzed as a sample of all residents living there.

Differences among the three control groups themselves were insignificant for most variables. The following discussion thus focuses on the contrast between TIR buildings and the three control groups taken together. Thus it examines the combined effect of tenants' participation and community work.

FACTORS AFFECTING PARTICIPATION IN THE PROGRAM

The first question considered was whether TIR buildings differ from the control groups with regard to characteristics and attitudes, which can explain both their selection by the organizers and the participation of the residents (Table 8.1).

Building characteristics. The level of maintenance is known to be affected by building design (Coleman, 1985; Werczberger & Ginsberg, 1987). Relevant factors include characteristics that facilitate the organization of the residents and the surveillance of the semipublic areas. In our sample, the number of entrances and thus of dwellings was significantly lower in TIR buildings than in the control groups (Table 8.1). Thus it seems that residents' organization is easier and participation more likely in small buildings with fewer tenants who know each other.

Socioeconomic characteristics. Participation in the TIR program requires motivation, organizational skills, social integration,and financial resources. From the literature,we know that demand for housing quality and maintenance increases with income (Mendelsohn, 1977). Moreover, owner-occupiers tend to maintain their property better than landlords, who consider only profits (Galster, 1983; Mayer, 1981), or renters, who enjoy the effects of maintenance only as long as they live in the dwelling. Home owners are thus also more likely than renters to participate in the management of condominiums (Ditkovsky & van Vliet--, 1985).

Respondents living in TIR buildings are indeed better off, as they are more likely to own cars, in Israel an important indicator of socioeconomic status (Table 8.1). A larger proportion are married, a status that facilitates participation in committee duties, yet they do not significantly differ with regard to other relevant characteristics, such as ownership, age, employment, occupation, crowding, and dependency on welfare (Werczberger & Ginsberg, 1987). This lack of relationship may be due to the relative homogeneity of the sample population. On the other hand, compared with the control groups, participating buildings have more new tenants, who are probably socially less integrated. Extended exposure to slum conditions seems to breed indifference to the physical environment, reducing interest in the program.

Attitudes. There are considerable differences between the two samples with regard to attitudes, which are indicators of alienation and might act as obstacles to organization (Table 8.1). Thus in TIR buildings tenants feel more at home in the semipublic areas. They complain less about troublemakers and helplessness with regard to what is happening in and around the building. Differences in preferences regarding the desirable level of maintenance complicate management and put a premium on social homogeneity (Mittelbach & Evin, 1975). In our sample, respondents in TIR buildings consider their neighbors to be significantly less heterogeneous with regard to origin and age (Table 8.1) than do the tenants in the control groups. There are also fewer who plan to leave the area because of dissatisfaction with the neighborhood.

EFFECTIVENESS OF THE PROGRAM

The main question for the study was whether or not the TIR program achieved its objectives. To find out, we looked for differences between TIR buildings and the control groups, which can be interpreted as the effect of participation on the relevant variables. Four goals were considered: organization of building committees, changes in attitudes with regard to upkeep, privatization of ownership, and improved maintenance. Note that a cross-sectional analysis cannot show the direction of causal relationships, that is, whether participation has led to positive attitudes or behavior or whether these preceded the implementation of

TABLE 8.1

Differences Between Building Groups:

Social and Physical Characteristics and Attitudes

Variables	F-Ratio[a]	t-Value[b]
Maintenance quality	(.00)	(.00)
Lighting condition	3.09	1.99
Number of entrances	4.16	3.37
Length of residence	2.94	2.07
No car	3.14	2.56
Percentage married	(1.44)	2.00
Satisfaction with dwelling	3.90	3.41
Satisfaction with building	(1.55)	1.85
Home territory	2.26	2.57
Knows neighbors	2.21	.81
Helplessness	5.09	2.52
Troublemakers	(1.05)	1.74
Similarity in origin	3.09	2.46
Similarity in age	1.93	2.13
Plans for future	2.91	(1.16)
Committee responsible	2.29	(.30)
Satisfaction with committee	5.27	2.56
Payment of dues	3.27	2.50

NOTE: The differences were analyzed using one-way ANOVA. Only variables for which the difference was significant level of $\alpha \leq .05$ are included.

a. The F-ratio was calculated for the four building groups; $df = 3$ for the between-group variance and 159 for within-group variance. An F-ratio of 2.67 is significant at the 5% level, and of 2.08 at the 10% level.

b. The t-value refers to the pooled variance estimate of the contrast comparing TIR buildings with all other buildings. The difference is significant at the 10% level for $t = 1.645$ and at the 5% level for $t = 1.98$.

the program and thus contributed to participation. Interpretation of the findings is thus inherently subjective.

Building committees. Maintenance of the shared areas in a multifamily building by the residents requires the continuing activity of a building committee, which was a precondition for participation in the TIR program. In the two types of buildings in which organizers had been operating, 80.7% had active committees, compared with only 43.7% in the remainder ($\alpha < .01$). Moreover, in these buildings there were significantly more residents who considered the committees responsible for maintenance, who were satisfied with the committees' activities, and who paid their dues. The results for TIR buildings were the same as the findings for those buildings whose tenants had also been exposed to the community workers but had not joined the program. Thus community work rather than participation seems to be responsible for these changes.

Residents' attitudes. The second objective of the program was improvement of attitudes regarding the housing environment and in particular regarding its maintenance. Residents of TIR buildings are, compared with the control groups, more satisfied with their dwellings and to some extent also with their buildings (Table 8.1). As already pointed out, we cannot know whether this is a cause or a result of their participation. In any case, there was no difference between the two groups in the ways they evaluated attitudes and behavior of their neighbors with regard to maintenance, in my judgment a good indicator of their own attitudes.

Privatization of ownership. The housing corporation expected that active involvement in renovation would stimulate the purchase of dwellings by tenants. One year after completion of the renovations, no significant difference was found between the samples with regard to the rate of ownership. This may be temporary, as 81.8% of the TIR renters expressed intentions to purchase their dwellings in the future, significantly more than the 67.8% of those in the control groups with such intentions ($\alpha < .02$).

Maintenance quality. The final question was whether the TIR program at least contributed to the privatization of upkeep. According to the data, there were no significant differences in maintenance quality between TIR buildings and the control groups.[2] Also from that point of view the program was not exactly a success, as participation apparently failed to have a direct effect on the quality of ongoing maintenance.

There remains the possibility of indirect effects. This would mean that participation modifies the influence on upkeep of other factors associated with demand for maintenance. Note that when the social composition of the residents offsets the effect of other factors, the resulting maintenance quality may, nevertheless, not be related to participation. For an analysis of these indirect effects, the simple correlation between maintenance quality and the explanatory variables was calculated separately for TIR buildings and for control groups (Table 8.2). The difference in sign or magnitude of the coefficients is then interpreted as an indirect effect of the TIR program.

TABLE 8.2

Zero-Order Differences Between Maintenance Quality and
the Independent Variables

Variables	TIR Buildings	All Other Buildings	Significance of Difference
Household size	.196	.369	—
Number per room	(.039)	.333	.069
Length of residence	.242	(.059)	—
Building satisfaction	.300	.364	—
Dwelling satisfaction	.162	.292	—
Plans for future	.184	.478	.049
Plans for children	(.118)	.292	—
Home territory	.284	.220	—
Troublemakers	(−.090)	.352	.101
Nobody cares	(−.129)	.279	—
Importance of maintenance	(.063)	−.241	.069
Reaction to litter	−.174	(−.159)	—
Committee responsible	(.148)	−.292	.008
Satisfaction with committee	.223	.008	—
Payment of dues	−.176	(.032)	—
Satisfaction with renovation	.188	—	—
Improvement in building	.299	—	—
Satisfaction with assistance	.262	—	—

NOTE: Only variables for which the difference was significant level of $\alpha > .95$ are included. Coefficients in parentheses are not significant.

The resulting differences were significant for only five variables: persons per room (crowding), plans for the future, presence of troublemakers, importance of maintenance, and the perception of the committee as responsible for upkeep. For each variable, the coefficient was significant only in the control group. The main effect of participation was apparently to reduce or neutralize the negative influence of certain characteristics or attitudes that tend to make maintenance or organization more difficult. Examples are the number of persons per room, which is related to the number of children, and negative attitudes with regard to the neighborhood and maintenance.

CONCLUSIONS

The TIR program attempted not only to improve the physical condition of participating buildings but also to promote the privatization of maintenance and ownership in public housing. The program itself was based on a temporary but significant privatization of renovation, which in the past had been considered the sole responsibility of the housing corporation.

The effects of the program were at best mixed. Renovation meant first of all a significant improvement in the condition of the shared areas. Residents in TIR buildings are also more likely to consider purchasing their dwellings, and their attitudes concerning the building committee seem to have improved. Yet attitudes regarding neither ongoing upkeep nor its quality appear to have been affected by participation in the program.

Nevertheless, residents in TIR buildings seem to have higher demand for housing and maintenance quality. This explains their willingness to participate in the program, yet it apparently did not sufficiently increase their motivation or ability to buy their dwellings or to carry out satisfactory ongoing maintenance after the completion of renovation.

Participation in the TIR program thus did not lead to the expected change in attitudes and behavior the sponsors had hope for. This should not be surprising. First of all, the contribution of the housing corporation was limited to assistance during the renovation process. Moreover, although purchase and improved upkeep were explicitly mentioned as goals of the program, they did not appear on the agenda of the organizers. Thus the assistance ended with the completion of the renovation—that is, at the moment when the real problems of maintenance began. Nor were organizers asked to bring up the sale of the apartments in discussions with the tenants. The residents, in turn, were motivated primarily by the physical improvement generated by the renovation and by the assistance provided by the housing corporation. They were not concerned with the goals of improved ongoing maintenance or privatization of ownership underlying the program. In conclusion, it seems that residents' involvement in a housing program may make a contribution to social objectives, including privatization, but only if these are also shared by the participating population.

NOTES

1. For a more extensive discussion of Israeli housing policy, see Werczberger (1988).

2. The exception was the condition of lighting and wiring, which was clearly better in renovated buildings. This part of the shared property is probably less dependent on

ongoing maintenance than other areas, and is thus less liable to deteriorate within one year after renovation.

REFERENCES

Burns, L. S. (1983). Self-help housing: An evaluation of outcomes. *Urban Studies, 20,* 299-309.

Carmon, N., & Gavrieli, T. (1987). Improving housing by conventional versus self-help methods: Evidence from Israel. *Urban Studies, 24,* 324-332.

Coleman, A. (1985). The social consequences of housing design. In B. Robson (Ed.), *Managing the city.* Totowa, NJ: Barnes.

Ditkovsky, O., & van Vliet--, W. (1985). Housing tenure and participation in resident and neighborhood committees. *Ekistics, 307,* 345-348.

Galster, C. G. (1983). Empirical evidence on cross-tenure differences in home maintenance conditions. *Land Economics, 53,* 107-113.

Galster, C. G., & Hesser, G. W. (1982). The social neighborhood: An unspecified factor in homeowner maintenance. *Urban Affairs Quarterly, 18,* 235-254.

Glazer, N. (1983). Towards a self-service society? *Public Interest, 70-71,* 66-90.

Katan, Y. (1974). Client participation in welfare services in Israel. *Social Security, 6-7,* 50-63. (in Hebrew)

Le Grand, J., & Robinson, R. (1984). *Privatization and the welfare state.* London: G. Allen.

Mayer, N. (1981). Rehabilitation decisions in rental housing. *Journal of Urban Economics, 10,* 76-94.

Mendelsohn, R. (1977). Empirical evidence on home improvement. *Journal of Urban Economics, 4,* 459-468.

Mittelbach, F. G., & Evin, J. 1975). Condominium housing: Some social and economic implications. *Journal of Sociology and Social Welfare, 3,* 170-181.

Oxman, R., & Carmon, N. (1986). Responsive public housing: An alternative for low-income families. *Environment and Behavior, 18,* 258-270.

Rosener, J. B. (1978). Citizen participation: Can we measure its effectiveness? *Public Administration Review, 38,* 457-463.

Sagiv, M. (1986, December). *The development of the rental housing market and the construction of rental housing in Israel.* Jerusalem: Ministry of Construction and Housing. (in Hebrew)

Saunders, P. (1988). The meaning of "home" in contemporary English culture. *Housing Studies, 4,* 177-192.

Silver, H., McDonald, J., & Vitiz, R. J. (1985). Selling public housing: The methods and motivations. *Journal of Housing, 42,* 213-221.

Ward, P. M. (Ed.). (1982). *Self-help housing: A critique.* London: Mansell.

Werczberger, E. (1988). The experience with rent control in Israel: From rental housing to condominiums. *Journal of Real Estate Finance and Economics, 1,* 277-293.

Werczberger, E., & Ginsberg, Y. (1987). The renovation of the shared areas in residential building in renewal neighborhoods. *Housing Studies, 2,* 192-202.

Part IV

Housing Finance

Introduction

MICHAEL A. STEGMAN

IN HIS TREATISE ON the sociology of housing tenure, Kemeny (1981) shows that postwar British, American, and Australian housing policies have been based on the assumption that home ownership is "inherently desirable and naturally superior to other forms of tenure, and that given accessibility and adequate resources, all households would choose to own" (p. 3). Unquestionably, policies promoting home ownership in each of these countries have been very successful, with at least 60% of all households owning their homes. In addition, home ownership rates within each country are higher among those in higher socioeconomic groups. In Britain, for example, about 91% of professional, as opposed to 21% of manual workers, are owner-occupiers. According to Kemeny, in the United States, 81% of high-income earners are home owners, compared with just 48% of low-income earners, while in Australia, about 80% of high-income earners are home owners, as against about 55% of low-income earners (p. 4).

The British, American, Australian, and other English-speaking countries' postwar housing experiences give rise to three hypotheses regarding housing tenure that Kemeny believes to be, "at best, highly questionable and, at worst, demonstrably erroneous." These are as follows:

(1) Home ownership and high socioeconomic status are necessarily positively correlated; in general, the rich tend to own and the poor tend to rent.

(2) Since low-income earners are unable to buy their homes (even though they invariably desire to do so), the extent of home ownership can be used as a measure of the material standard of living of a country. It therefore follows that countries with high home ownership rates will tend

to have high material standards of living, while countries with low home ownership rates will tend to have low material standards of living.

(3) If it is argued that rich industrial societies have higher home ownership rates than poorer industrial societies, then it also follows that as countries increase their material standard of living the home ownership rate can be expected to rise. This suggests that, as the standard of living rises, more and more households will be able to afford to buy their homes, until all families will be home owners.

Although the Anglo-Saxon experience tends to support these hypotheses, Kemeny shows that the relationship between housing tenure and national standard of social well-being is not nearly as strong in other European societies. Thus, for example, despite the fact that Sweden, West Germany, the Netherlands, and Switzerland are among the richest countries in the world, they each have home ownership rates of one-third or less. Moreover, these same countries have more highly developed social welfare systems than do most English-speaking nations, the implication being that the "high cost of house purchases . . . acts as a deterrent to increasing taxation levels as a prerequisite for extending social welfare" (Kemeny, 1981, p. 56).

> Thus, for example, in terms of the percentage of income paid in direct taxes after deductions in 1976/77, Sweden and West Germany, with 34.4 percent and 27.8 percent respectively, are well ahead of Britain, Canada, France and the USA, with all around 20 percent. In terms of social service expenditure as a percentage of Gross National Product the Netherlands (20 percent), Sweden (19 percent), and West Germany (17 percent) are ahead of France (14.5 percent), the United Kingdom (14 percent) and Canada (13.5 percent), while New Zealand (11.5 percent), the USA (9.5 percent) and Australia (9 percent) spend the least. (p. 55)

It is also the case that most government assistance to home owners is indirect, generally coming in the form of various kinds of tax deductibility rather than in direct payments. These subsidies, Kemeny points out, "are not typically defined as 'welfare' " (p. 75). Yet, they

> represent transfers of wealth from tenants to owners which do not carry the stigma associated with the payments of subsidies to 'public housing'. The importance of this lies in the fact that, if homeowners can be indirectly subsidized while, say, public housing tenants can be directly subsidized, it both legitimates homeownership and stigmatizes non-owners. It therefore readily permits homeowners to be defined as "paying their way" in

housing, as against those who are unable to do so and must be "supported by the State." (p. 75)

In short, Kemeny holds that, rather than being an inherently superior tenure form and the unplanned by-product of sustained high rates of postwar economic growth, the preference for and high rates of home ownership that have been achieved in many English-speaking countries during periods of sustained economic growth are the result of conscious, culture-based, and costly government policies that have distorted housing choices and affected tenure patterns in undesirable ways. It is appropriate, therefore, to ask what national housing policies in these countries should be now that their economic futures are far less certain than they have been at any time in the past 40 years.

Against the backdrop of a common culture of home ownership, in the chapters that follow, Roistacher, Carter, and Crook provide valuable insights into future housing finance policies, respectively, in the United States, where the home ownership rate has fallen for the first time in 40 years; Great Britain, where home ownership grew to its highest level ever during the 1980s, thanks largely to a massive "tenant right-to-buy" program, and has now peaked; and Australia, where financial deregulation threatens to prevent young people from buying a house in a country in which home ownership is a "cultural obsession."

Roistacher's correct assessment of the housing policy legacies of the Reagan era is that "the overall system of federal subsidy is certainly more regressive than before, as a result of spending cuts and reductions in the progressivity of the income tax." As a result, states and localities "are under increasing pressure to address low- and moderate-income housing needs through their own resources." The fact that states have resigned themselves to the federal government's retreat from its historic responsibility to support low-income housing activities is reflected in the results of one national survey of state initiatives. Roistacher notes that the survey showed that "of 153 identified programs in which states expended their own funds, 31% were enacted between 1980 and 1984 and 34% were initiated in 1985 or 1986."

She also demonstrates that "Reagan has also created the image of the deficit as the number one national economic problem," which she argues is "a debatable proposition." It is especially debatable in light of the sizable "peace dividend" that many (particularly, liberal) political analysts believe should result from the sizable reduction in the levels of U.S. armed forces that have been made possible by *glasnost'* and the recent dramatic events in Eastern Europe. Although the current policy of the Bush administration is to deny that any such peace dividend is possible, the major U.S. public policy debate of the 1990s, which will have major implications for low-income housing and other areas of domestic policy, is whether the substantial savings made possible by a reduction in national defense spending will be used for deficit reduction and tax cuts or to support housing and other social welfare programs that have been "starved by eight years of Ronald Reagan" (Evans & Novak, 1989).

Like the United States, Australia has witnessed substantial financial deregulation during the 1980s, including the entry of foreign banks, mergers between building societies and banks, legislation for building societies that encourages greater diversification, the removal of many asset controls on the banking sector, and the deregulation of interest rates on savings bank home loans. As one might expect, in a nation in which about 70% of all households own their homes, when home lending rates rise substantially faster than other interest rates, what Robert Carter calls "a countervailing response" becomes an urgent political necessity.

Unlike Roistacher's broad-based policy analysis, Carter's essay focuses on recent developments in Australia's secondary mortgage market and the creation of and attempts to institutionalize the use of a single new mortgage instrument—the price level adjusted mortgage (PLAM)—that is designed to give moderate-income households whose real incomes are likely to be stable or increasing over time continued access to the home financing system without government subsidy.

Under a conventional fixed-rate mortgage, the real value of loan repayments are highest at the beginning of the loan term and, because of inflation, lowest at the end. With a PLAM, "the time path of repayments is shifted to one in which the real value of house repayments is steady over time, rather than continuously declining." According to Carter, the interest rate on a PLAM consists of a real interest rate that is set at 3% per year for the lowest income borrowers. The initial loan repayments are set at 25% of gross household income, and are adjusted each year for changes in the inflation rate. Should a home buyer's income not increase at least as much as the inflation rate during the year, he or she can petition to have the 25% rule restored and extend the term of the loan. Carter smoothly guides the reader through all the necessary technicalities of inflation-adjusted mortgage loans and, more important, assesses the cultural and institutional responses to this new form of home financing. Among other things, he shows that, when it comes to consumer and lender acceptance, technology transfer in the mortgage lending sphere may take as much or more time to accomplish than it does in the goods-producing sector.

Part IV concludes with Crook's chapter on housing finance in Great Britain. Rather than reflecting on the housing policy revolution of the first two Thatcher governments, which saw more than a million council housing units bought by sitting tenants at discounts of up to 70%, Crook focuses on the plight and future possibilities of the British private rental sector. Unlike the United States, for example, where about a third of all households live in privately owned rental housing, this is the case for less than 11% of all British households. Also in contrast to the United States, the private rental sector in Great Britain has the greatest concentration of substandard conditions—due primarily, though not entirely, to decades of stringent rent controls—and houses the poorest of the poor.

Based on the premise that the rejuvenation of the private rental sector is necessary to Britain's broad economic recovery, the Thatcher government has announced new plans to deregulate all new tenancies in the existing privately owned stock, end local council construction of new rental housing, transfer local authority rental estates to a range of private landlords, and encourage the construction of new private rental housing through a combination of mixed private and public funding. Crook argues that the government's efforts to deregulate and, thereby, reinvigorate the private rental sector will not work because of the limited rent-paying abilities of those who rely on this sector for housing. "Unfortunately," he says, "deregulation as envisaged conflicts with government policy to cut public spending on housing and to maintain fiscal support for owner occupation, which reduces demand for private renting." In essence, then, Crook brings us full circle in the policy debate. As Kemeny (1981) has shown, it is not a simple matter to restore balance to national housing policies that have distorted investment decisions, housing choices, and tenure patterns through disproportionately large commitments to home ownership.

REFERENCES

Evans, R., & Novak, R. (1989, December 22). Gramm divides peace dividend. *Boston Globe.*

Kemeny, J. (1981). *The myth of homeownership: Private versus public choices in housing tenure.* London: Routledge & Kegan Paul.

9

Housing Finance and Housing Policy in the United States: Legacies of the Reagan Era

ELIZABETH A. ROISTACHER

RONALD REAGAN'S YEARS AS PRESIDENT of the United States have left a distinct imprint on the federal government: Domestic spending has been sharply constrained as discretionary programs have been cut and entitlements have been slowed by imposing restrictive eligibility criteria (U.S. Executive Office of the President, 1989a). In addition, the tax system has been restructured, and policy discussions invariably include the need for more state and local government involvement and for greater private sector participation. In no area of domestic spending have these changes been felt as severely as in low-income housing.

The new Republican president, George Bush, is approaching the end of his first year in office. While it remains to be seen how he and Secretary of Housing and Urban Development Jack Kemp will attempt to guide housing policy, Kemp's market-oriented economic philosophy is well known. Substantive discussion of low-income housing policy has recently been sidetracked by revelations of serious political manipulation of HUD programs as well as outright fraud. These scandals could be used as justification for further dismantling of federal housing programs. Any policy outcomes, of course, depend not only on the goals of the administration but on those of the Congress, both houses of which are controlled by the Democrats.

AUTHOR'S NOTE: I gratefully acknowledge the indispensable research assistance of Rachel Kaufman and financial support from the Robert F. Wagner, Sr., Institute of Urban Public Policy, a PSC-CUNY Faculty Research Award, and a Queens College Presidential Research Award.

This chapter reviews U.S. housing policy, changes in that policy during the Reagan years, and, to the extent possible, future directions. It also examines responses by state and local government to these changes. The chapter concludes with an assessment of the consequences with respect to social and economic welfare.

BACKGROUND: THE NATURE
OF FEDERAL HOUSING POLICY

In the United States, housing finance tends to be narrowly defined as mortgage finance. This chapter takes a broader view, considering the wide array of government financial incentives to housing. In this context, financial deregulation, tax reform, and government spending for housing are relevant.

The federal role in housing began in earnest in 1937, with the creation of the public housing program. Justification of this legislation was first and foremost as a stimulus to employment in the construction industry. Promoting safe and sanitary housing for low-income people was a second objective (see Bellush & Hausknecht, 1967; Sternlieb & Listokin, 1987). The federal government provided the capital subsidies to local housing authorities, which were expected to construct the housing and charge rents that would cover operation and maintenance. Thus there was a floor, as well as a ceiling, on the incomes of public housing tenants to assure that rents could be covered. Like Western European social housing, it was very much a program for the working class, the "deserving poor." However, with the introduction of federal operating subsidies in 1969, public housing has become a houser of the nation's very poor families.

THE GREAT DEBATE:
PRODUCTION SUBSIDIES
VERSUS TENANT SUBSIDIES

In the 1960s, federal policy shifted toward "assisted housing"—privately owned and operated but in receipt of a variety of federal subsidies, including below-market financing, mortgage insurance, tax benefits, and rental assistance for low-income occupants.

After a 1973 moratorium on federal housing programs imposed by President Richard Nixon, and a major National Housing Policy Review (1974), the idea that supply-side subsidies were less efficient than subsidies to tenants became the dominant theme in housing policy debates. New construction has been found to cost roughly twice as much in government subsidy as rent support for a tenant living in an existing dwelling (U.S. Executive Office of the President, 1989a). It is also argued that tenant-based subsidies give households the "freedom" to choose a dwelling with the characteristics that they want, so any given amount of subsidy is likely to have greater value than if the government "imposes" a unit on a household. (This argument assumes that households have a real choice.)

The National Housing Policy Review resulted in the passage of Section 8 of the Housing and Community Development Act of 1974, creating two new programs: Section 8 new construction/substantial rehabilitation was much akin to former supply-side programs; Section 8 existing housing was a more innovative demand-side initiative in which low-income tenants were given participation certificates that could be used to pay for the housing they currently occupied or to shop for other housing in the existing stock.

By the time Section 8 was fully under way, Jimmy Carter was president. During his administration production subsidies—heavily supported by the construction lobby—continued to dominate actual activity. His last budget estimated 1981 housing activity at about 250,000 units, with more than 50% of these units being for new construction or substantial rehabilitation, the rest going either to rent certificates or to moderate rehabilitation, a subsequent addition to Section 8 legislation (U.S. Executive Office of the President, 1981a).

HOME OWNERSHIP SUPPORT:
THE HIDDEN SUBSIDIES

At the same time that the vocal housing debate focused on whether low-income housing programs should be production oriented or tenant oriented, a silent housing policy to assist homeowners continued to be carried out through the tax code and various federal housing finance agencies. Tax deductions for homeowners' mortgages and property taxes and special capital gains treatment of owner-occupied residences are the most important part of this agenda. In addition, the special tax treatment of thrift institutions (those specializing in home mortgages) and the creation of federal mortgage insurance programs and federal secondary mortgage market agencies were all a part of the home ownership support system.

The federal budget—for both outlays and tax expenditures (revenues forgone through special tax treatment)—gives a good deal of insight into the relative importance of low-income rental housing policy versus home ownership policies (see Table 9.1) and also provides a basis for examining changes in policy over time. In 1980, home ownership tax

TABLE 9.1

U.S. Tax Expenditures, Budget Outlays, and Budget Authority for
Housing, 1980 and 1988 (in millions of dollars)

	1980	1988	% Change
Tax expenditures			
owner-occupied housing			
deductibility of mortgage interest	15,615	33,675	115.6
deductibility of property tax	7,310	10,100	38.2
exclusion of interest on state and local			
housing bonds	447	1,765	294.9
deferral of capital gains on home sales	1,010	3,700	266.3
exclusion of capital gains on home sales,			
persons aged 55 or over	535	2,940	449.5
excess bad debt reserve of financial			
institutions	470	80	−83.0
total owner-occupied tax expenditures	25,387	52,260	105.9
renter-occupied housing[a]			
exclusion of interest for state and local			
housing bonds	310	1,235	298.4
accelerated depreciation	385	300	−22.1
expensing of construction period interest			
and taxes	695	—	−100.0
five-year amortization for rehabilitation of			
low- and moderate-income housing	15	45	200.0
credit for low-income housing investment	—	160	—
total rental tax expenditures	1,405	1,740	23.8
total tax expenditures for housing	26,792	54,000	101.6
Budget outlays for subsidized housing			
and public housing	5,376	12,711	136.4
Budget authority for subsidized housing			
and public housing	27,536	8,605	−68.8

SOURCE: U.S. Executive Office of the President (1981b, Special Analysis G, Table G-1; 1989b, Special Analysis G, Table G-1).
a. The 1988 tax expenditure budget includes $1.1 billion for a rental housing exemption to passive loss restrictions introduced in the Tax Reform Act of 1986. A redefinition of the normal tax standard created this tax expenditure. There is no equivalent item in the 1980 tax expenditure budget.

expenditures exceeded $25 billion; those to encourage rental housing were just over $1 billion. Budget outlays for low-income housing programs were just over $5 billion.

Budget outlays represent one year's support, including the current year's expenditures for previous program commitments and for new commitments. Budget authority, on the other hand, represents long-run commitments to housing programs that result from new commitments in the current fiscal year.[1] In 1980, budget authority was $27 billion, more than five times the level of current outlays.

When Ronald Reagan took office, the U.S. social housing sector assisted 4.2 million households (5% of the nation's 83 million households); 1.2 million lived in public housing, 2 million lived in HUD-assisted housing or received Section 8 certificates to pay rents in private housing, and another million households, many of whom are moderate-income home owners, received assistance under programs sponsored by the Farmers Home Administration (FmHA), an agency that supports rural housing activities.

THE REAGAN RECORD

Reagan reshaped housing policy not only through substantial restructuring of low-income housing programs but also through certain aspects of tax reform. Despite a major revamping of the tax system, however, tax preferences for home owners remain virtually untouched. During the Reagan years, the thrift industry underwent substantial deregulation. Adverse consequences of deregulation were apparent throughout the Reagan years, although no major action was taken to try to address the shortcomings of deregulation until Reagan left office.

LOW-INCOME HOUSING:
THE VOUCHER PROGRAM

The major focus of the Reagan administration with respect to low-income housing programs was elimination of production subsidies and the development of a voucher program that gives assistance directly to tenants. The voucher program operated on an experimental basis in its early years and received final congressional approval in 1987. The voucher is a refinement of the Section 8 rent certificate. Under the voucher program, the household is paid the difference between a specified "fair market rent" and 30% of adjusted income. The household may spend more for housing than the fair market rent, although additional rent dollars are not subsidized. If the household spends less than fair market rent, it may keep the difference between the voucher and the actual rent. Under Section 8, a household could pay no more for its housing than the specified fair market rent. If the household found housing below the fair market rent, its federal support would be based on actual rent. Thus there was little incentive to shop for housing below the fair market rent.

While the voucher is touted because it allows households to spend more of their own money if they want better housing, a number of program changes introduced during the Reagan administration made this an empty promise. Households are now asked to contribute 30% of income to housing rather than the 25% they did under Section 8, and program eligibility was lowered from 80% to 50% of area median income. As of the end of 1988, 175,000 vouchers had been issued to local housing authorities. HUD studies indicate that roughly 60% of the time a household will return a voucher to the local housing authority unused, which suggests that it is often difficult for recipients to find qualifying housing. (The voucher is then issued to another recipient.)[2]

The consequences of the Reagan administration's efforts to phase out production subsidies is evident from the dramatic decline in budget authority for low-income housing programs (see Table 9.1): It fell by 69% in current dollars, from $27 billion in 1980 to $8.6 billion in 1988; in real (inflation-adjusted) terms the decline was 78%.[3] Housing programs have been cut less than the Reagan administration would have liked. A number of administration budget requests called for a moratorium on federal housing programs. In virtually every year, the Congress increased allocations beyond administration requests.[4] Reagan's last budget submission, for fiscal year 1990, proposed funding for another 132,000 households, including 120,000 vouchers (20,000 for rural housing programs) and just under 12,000 production subsidies.

To the extent that shifting from production subsidies to rent subsidies was justified as a mechanism to allow federal dollars to go farther, these dollars were not used to increase the rate at which the government expanded support to low-income households. During the Reagan years, the number of assisted households grew by nearly 30%—from 4.2 million to 5.4 million (U.S. Executive Office of the President 1989a). The annual increase of about 150,0000 households was roughly half of what it had been in the years prior to Reagan. Moreover, a portion of the increase reflected commitments made by previous administrations. Of 13 million eligible renter households, roughly a third are receiving support.

THE PROBLEM OF EXPIRING SUBSIDIES

Many past federal commitments are reaching a point of possible expiration. Potentially, somewhere between one and two million units developed as low-income housing could be "privatized" or disappear from the low-income housing inventory over the next decade. When private developers contract with HUD for subsidies—involving mortgage insurance, below-interest mortgages, or rent subsidies—the federal government in turn limits rents, profits, and occupancy during the contract period. Older production programs allow for prepayment of mortgages and a cancellation of the regulatory agreement after 20 years. By 1993, some 600,000 units in HUD and FmHA programs will be eligible for mortgage prepayment. Another one million federally supported low-income units will have their rent subsidy contracts expire or be eligible for termination (National Housing Task Force, 1988). One study estimates that 38% of HUD's existing inventory is likely to opt for prepayment and that another 43% is likely to default (National Low Income Housing Preservation Commission, 1988; see also Chapter 6, this volume).

Legislation has placed what is virtually a moratorium on mortgage prepayments in HUD programs, and existing appropriations have so far been adequate to extend rent contracts that have expired. However, without a major federal initiative, it is clear that hundreds of thousands of families will face sharply rising rents or displacement over the next decade. The National Low Income Housing Preservation Commission (1988) has estimated that to protect HUD's existing low-income housing inventory will require an additional $11.3 billion over 15 years, and additional funding would be required to protect FmHA units.

TAX REFORM AND
THE LOW-INCOME HOUSING CREDIT

Special tax treatment of home ownership has been part of the federal tax code from its earliest days. In addition to deductibility of mortgage interest and local property taxes, capital gains are deferred upon the sale of a home if they are reinvested in another home; the elderly are allowed a capital gains exclusion of up to $125,000. (See Table 9.1 for the costs of these to the federal treasury.) Despite the fact that the Reagan administration considered tax reform as one of its major accomplishments, these home owners' tax preferences remained untouched. Tax reform has, however, affected their value by lowering marginal tax rates and therefore the value of deductions to higher-income households.

The first tax legislation during the Reagan years, the Economic Recovery Tax Act (ERTA) of 1981, was designed to spur investment and to begin rolling back marginal tax rates. (The only change in marginal rates, however, was a decline in the top bracket from 70% to

50%.) The major stimulus to investment was through the Accelerated Cost Recovery System—and rental housing investment was one of the beneficiaries of this provision. For all rental housing the minimum depreciation period was lowered from an average of 30 to 15 years. Low-income rental housing received further benefit through a 1-year write-off of construction period interest and taxes and a more generous method of depreciation. It has been estimated that the net effect of the 1981 act was to lower required rental income for low-income housing by anywhere from 11% to 23% below what would be required for comparable housing in the absence of the low-income benefits (Villani, 1987). The costs to the federal treasury associated with ERTA were so great that Congress passed several acts to roll back tax benefits, and rental housing investment incentives were reduced.

The last major tax act of the Reagan years, the Tax Reform Act (TRA) of 1986, was the heart of Reagan's tax reform agenda. It was designed to be revenue-neutral, reducing tax brackets and progressivity at the same time that it broadened the tax base by eliminating deductions and other tax preferences—but not those for home owners. This act put added curbs on rental housing investment by lengthening depreciable lives (to 27.5 years), by raising the rate of taxation of capital gains, and by reducing opportunities for tax shelters through increases in minimum taxes and restrictions on passive losses. Passive losses are tax write-offs that accrue to investors who do not have direct involvement in a project; the "syndication" of tax shelters to passive investors had been the major stimulus to attracting private funds to low-income housing.

Because the main portions of the Tax Reform Act were so devastating to investment in low-income housing, Congress added a tax credit for low-income housing. Tax credits of up to 70% in present value are available to qualifying low-income projects. Credits are significantly lower (30% in present value), however, if the assisted project uses any other form of federal subsidy, most notably federally tax-exempt housing bonds issued by state and local governments (U.S. Executive Office of the President, 1989b). The tax credit was authorized through 1989; as of this writing, Congress is deliberating its renewal.

The tax credit generally proved unworkable unless substantial other subsidies were also available. Moreover, the potential market for tax credits (taxpayers with high marginal tax rates and a need to shelter income) was decimated by the TRA's restrictions on passive losses. Technical details designed to target the benefits to low-income families often proved so onerous as to make projects and credits unmarketable. Finally, syndication costs to attract high-income investors are quite high, adding, according to one experienced nonprofit group, up to 20% to development costs.

Despite its shortcomings, the tax credit program appears to be the only real federal support of low-income housing production. As such, despite its inefficiencies, housing advocates are supporting its renewal with some modifications to make the program more effective. From the U.S. Treasury's perspective, of course, the more effective the program, the more costly it is in forgone tax revenues.

THE SAVINGS AND LOAN CRISIS

In the early days of the Bush administration, the mounting failures in the nation's thrift industry demanded immediate government attention. Some history is essential to understanding the current crisis.[5]

Ever since the 1930s, thrift institutions (savings and loans and mutual savings banks) have specialized in mortgage finance in exchange for special regulatory and tax treatment. Thrifts were required to lend "long" on fixed-rate home mortgages. While not required to do so, they borrowed "short," predominantly in the form of retail deposits insured by a quasi-governmental agency, usually the Federal Savings and Loan Insurance Corporation (FSLIC).

This system worked well until the onset of inflation in the mid-1960s. When the cost of funds began to rise, the government protected thrifts by imposing deposit rate ceilings and raising the minimum denomination of the best available alternative investment, U.S. Treasury bills (White, 1989). As U.S. financial markets developed and depositors found new alternatives to thrift deposits (in particular, money market funds), high interest rates produced the problem of "disintermediation": Funds flowed out of thrift institutions into higher-yielding assets, and thrifts were forced to reduce mortgage lending.

By the early 1980s, thrifts were permitted to offer higher yields on certain deposits and to provide adjustable-rate mortgages. Nevertheless, they continued to suffer from a fundamental problem of maturity mismatch, paying high costs for short-term deposits but earning low yields on their portfolios of home mortgages, most of which were fixed-rate loans committed when rates were lower. By this time, deregulation had become a byword of government, and the financial sector became a prime target. Interest rate ceilings were completely eliminated and thrifts were permitted much riskier forms of lending. Other institutions were authorized to become mortgage lenders.

Deregulation, however, has been disastrous for thrifts for a combination of reasons: the continuing problem of maturity mismatch, the mounting losses on risky commercial investments, and growing losses on mortgage loans as a result of poor underwriting in depressed housing markets. Low capital requirements and no risk adjustment of deposit insurance premiums encouraged thrifts to take excessive risks. A combination of poor judgment and outright fraud has resulted in mounting industry losses. By the time Bush took office, it was estimated that the eventual cost to protect depositors would exceed $300 billion over a 30-year period.

Legislation passed soon after Bush entered office has restructured the thrift regulatory system. Capital requirements have been increased so that the owners of institutions have more at risk, and new constraints have been placed on thrifts' lending activity. It is estimated that the industry may shrink from 3,000 to 1,000 institutions as mergers and closures continue. The new plan is to be funded three-fourths by government and one-fourth by healthy thrifts (Hershey, 1989).

Despite the serious crisis that has confronted the thrift industry, financial deregulation has, for the most part, had positive consequences for mortgage markets. The thrift industry, despite all its problems, has contributed increasing volumes of funds to the mortgage market.

Perhaps the biggest problems for home owners are those associated with adjustable-rate mortgages. When these were first actively originated, some lenders offered artificially low "teaser" rates that put both lender and borrower in financial jeopardy as rates rose and, in some communities, house values fell. Problems with adjustable-rate mortgages have abated over time. Lenders have tended to eliminate teasers as a result of adverse experience and the imposition of tougher standards by secondary market makers. Regulators have introduced some standardization in the information that lenders provide consumers.

THE SHIFT TO
STATE AND LOCAL GOVERNMENTS

The cutbacks in federal housing assistance, tax reforms, and new strictures on housing bonds put states and localities under increasing pressure to address low- and moderate-income housing needs through their own resources. The problem of expiring subsidies on older federally supported housing is likely to add to state and local burdens. The most visible consequences of not taking action in many communities will be increasing homelessness. The hidden problem will be growing economic burdens on the poor, increasing doubling-up of families, and eroding housing conditions. For places like New York City, the inability to provide housing for moderate-income families threatens continued economic growth.

While it is difficult to summarize what is happening in 50 different states and hundreds or thousands of localities, there are a few general conclusions that can be drawn. First, virtually all states have some form of housing program, historically motivated by their ability to issue tax-exempt housing bonds. Second, state involvement in housing is accelerating (Sidor, 1986).

City governments have also become increasingly active in directly supporting housing. One recent survey of locally funded housing programs examined the activities of the 51 largest cities in the United States and found that nearly half of them are devoting some portion of locally generated dollars to affordable housing production and rehabilitation (Berenyi, 1989). Of the 26 cities that do not dedicate any local revenues to housing, 7 are restricted by state law from doing so, and the remaining 19 cited various other reasons: budget constraints, high vacancy rates, or more pressing other needs. New York City's level of commitment came to $102 per capita. The next highest spender on a per capita basis was Honolulu, Hawaii, at $74. Other major cities with severe housing problems, such as Boston, Los Angeles, San Francisco, and Washington, D.C., were spending significantly less ($5.75, $15.49, $10.48, and $12.38, respectively).

Because cities find it difficult to fund low- and moderate-income housing themselves, quite a few attempt to induce the private sector to provide such housing. For example, in San Francisco and Boston, downtown commercial projects are charged a per-square-foot fee that goes to a fund to subsidize affordable housing. In New York, tax abatements are being granted to developers who finance low-cost housing elsewhere in the city. Other communities have turned to inclusionary zoning that preconditions building of luxury housing on the provision of affordable housing. Others grant zoning bonuses to those who provide affordable housing units. The use of exactions to meet low-income housing needs is a form of privatization of housing policy, but one that stems more from the inability of localities to fund such activities themselves than from a philosophy about the desirability of private versus public activity.[6]

Local funding of low-income housing has become a political and moral necessity. Unfortunately, it is fiscally irrational. When local governments take on the burden of redistributing income, adverse fiscal consequences result. Taxpayers' willingness to tolerate tax burdens is a function of what they perceive they are receiving for their tax dollars. The typical taxpayer does not view income redistribution as a direct benefit; thus, as local tax burdens rise to finance income redistribution, the tax base will erode, as taxpaying households and firms with reasonable options relocate to tax jurisdictions with lower tax burdens (see Ladd & Bradbury, 1988). It is also argued—although without any strong empirical support—that local income redistribution policies will attract new residents in need of support, further driving up the local tax burden and incentive to tax flight by taxpayers. Thus declines in federal support for low-income people and their housing put cities on the horns of a dilemma: Either ignore the needs of the poor or place damaging fiscal burdens on taxpayers. The U.S. system of fiscal federalism, as it is practiced, contrasts sadly with the more centralized fiscal responsibility in such Western European countries as Sweden and the Netherlands (Roistacher, 1987a).

HOUSING POLICY AFTER REAGAN

Eight years of Reagan administration policies eroded federal support of low-income housing by more than 75%. Federal support of more than a million older low-income housing units is at risk of expiring over the next decade. Home owners' tax preferences have remained virtually intact despite major tax reform, and tax reform has resulted in reduced private incentives to produce rental housing. The low-income housing tax credit program, which was difficult to utilize, is about to expire.

Secretary of Housing and Urban Development Jack Kemp has an established record for being concerned about cities and the poor. His general approach is highly market oriented: tenant management of public housing, selling public housing to tenants (see Chapter 7), housing vouchers, and stimulating urban development through enterprise zone. But his ability to hone a housing policy is temporarily hindered as he copes with the unraveling of his agency as a result of the mismanagement at HUD during the Reagan years.

THE HUD SCANDALS AND BEYOND

During the Reagan years, Secretary Samuel R. Pierce, Jr., had perhaps the lowest profile in the cabinet. In 1989, after Pierce and Reagan had left office, a series of scandals materialized (see Waldman, 1989, for a detailed account). The first revelation concerned influence peddling, occurring mostly in the moderate rehabilitation program. Mod-rehab funds were originally allocated to localities on a formula basis, as were other Section 8 funds, but the program was dramatically cut and was folded into the secretary's discretionary fund. Staff recommendations and assessments of local governments on the advisability of projects were frequently overridden.

Another major problem at HUD developed in its coinsurance program. HUD turned to the private sector to help insure properties. However, private agents were able to profit from the program by overvaluing properties, which helped increase their fees. Even if some properties defaulted, the agents did well because total fees exceeded the costs they had to absorb as coinsurers. In addition, HUD turned to private agents to sell defaulted property. Some agents simply pocketed the funds for themselves.

The HUD scandals have temporarily overshadowed any substantive discussions of housing policy. However, they appear to be lending additional support to vouchers over production programs. The secretary and his advisers perceive the discretion associated with production programs as creating opportunities for political abuse (Cohn, 1989). This is a questionable conclusion: The moderate rehabilitation program became ripe for abuse because it was cut so much that it could be administered only on a discretionary basis. Problems of fraud appeared in those programs that had been "privatized," and cuts in staff made it difficult for HUD to monitor private agents.

HOMELESSNESS AND
THE FUTURE OF HOUSING POLICY

Over the Reagan years, homelessness emerged as a pressing national problem, and its aggravation is a part of the Reagan legacy. Estimates of the number of homeless nationally range from the administration's figure of 250,000 to the 3 million asserted by advocates for the homeless. Ironically, the problem has risen so high on the political horizon that the Bush administration will be forced to confront it. Even Reagan's last budget included an allocation for 2,000 units of homeless housing. Secretary Kemp took an immediate interest in the problem by touring homeless shelters, but he has yet to advocate a federal role in dealing with the problem.

Most federal support to shelter the homeless is emergency assistance from the Department of Health and Human Services. These emergency assistance funds cannot be used to produce housing; they can be used only to provide temporary shelter. The view of homelessness as a temporary condition, together with federal cutbacks in permanent housing, can result only in aggravated low-income housing problems and greater fiscal burdens on cities.

Any new program initiatives, however, are likely to be quite different from those of the past. The Reagan administration was very effective in shifting emphasis away from direct federal intervention. In particular, there is likely to be increased emphasis on public-private partnerships at the state and local levels, on nonprofit providers, and on tax credits rather than direct subsidies. Indeed, the tax credit program is very much tied into this new environment. Many projects are initiated by the nonprofit sector and receive substantial additional support from state and local government to make them feasible. Housing policy recommendations of the congressionally appointed National Housing Task Force put a heavy emphasis on this "defederalization" of federal housing policy.

In sum, the legacy of Reagan is not only cuts in federal support of housing and other social initiatives but also a change in the nature of future policy discussions. The Reagan administration slashed budget authority and tax expenditures for low-income housing but continued to let home owners' tax benefits grow. The overall system of federal subsidy is more regressive than before as a result of spending cuts and reductions in the progressivity of the income tax. The rhetoric of efficiency and deficits has been used to restructure the federal budget away from social policy and toward national defense, to shift appropriate federal income-redistributive responsibilities to lower levels of government, and to reduce social and economic welfare in the United States.

NOTES

1. Budget authority tends to be high relative to outlays if new commitments involve production-based subsidies, because they require spending over a long time horizon—anywhere from 15 to 30 years.

2. Unless otherwise cited, data on HUD programs in this section were provided by the Office of Policy Development and Research, U.S. Department of Housing and Urban Development.

3. The 1988 budget authority is significantly less than 1988 outlays; this is because new commitments are relatively short term (voucher contracts last only 5 years, compared with the 15- to 30-year commitments of production programs), and new commitments themselves account for only a small portion of current outlays.

4. See Zigas (1989) for a detailed discussion of Reagan's budget requests for housing.

5. See White (1989) for a detailed history of the crisis and recommendations for its resolution. See Roistacher (1987b) for a comparison with the experiences of the British building society industry.

6. See Alterman (1988) for an extensive treatment of these linkage policies.

REFERENCES

Alterman, R. (Ed.). (1988). *Private supply of public services.* New York: New York University Press.

Bellush, J., & Hausknecht, M. (1967). Urban renewal: An historical overview. In J. Bellush & M. Hausknecht (Eds.), *Urban renewal: People, politics, and planning* (pp. 3-16). Garden City, NY: Anchor.

Berenyi, E. B. (1989). *Locally funded housing programs in the United States: A survey of the 51 most populated cities.* New York: New School for Social Research, Community Development Research Center.

Cohn, B. (1989, August 21). Looking beyond the HUD scandal. *Newsweek*, p. 19.

Hershey, R. D., Jr. (1989, August 10). Bush signs saving legislation: Remaking of industry starts fast. *New York Times*, p. A1.

Ladd, H. F., & Bradbury, K. L. (1988). City taxes and property tax bases. *National Tax Journal, 41*(4), 503-523.

National Housing Policy Review. (1974). *Housing in the seventies.* Washington, DC: U.S. Department of Housing and Urban Development.

National Housing Task Force. (1988). *A decent place to live.* Washington, DC: Author.

National Low Income Housing Preservation Commission. (1988). *Preventing the disappearance of low income housing.* Washington, DC: National Corporation for Housing Partnerships.

Roistacher, E. A. (1987a). Housing and the welfare state in the United States and Western Europe. *Netherlands Journal of Housing and Environmental Research, 2*(2), 143-175.

Roistacher, E. A. (1987b). The rise of competitive mortgage markets in the United States and Britain. In W. van Vliet-- (Ed.), *Housing markets and policies under fiscal austerity.* Westport, CT: Greenwood.

Sidor, J. (1986). *State housing initiatives: A compendium.* Washington, DC: Council of State Community Affairs Agencies.

Sternlieb, G., & Listokin, D. (1987). A review of national housing policies. In P. D. Salins (Ed.), *Housing America's poor* (pp. 14-44). Chapel Hill: University of North Carolina Press.

U.S. Executive Office of the President. (1981a). *Budget of the United States government fiscal year 1982.* Washington, DC: Government Printing Office.

U.S. Executive Office of the President. (1981b). *Special analyses: Budget of the United States government fiscal year 1982.* Washington, DC: Government Printing Office.

U.S. Executive Office of the President. (1989a). *Budget of the United States government fiscal year 1990.* Washington, DC: Government Printing Office.

U.S. Executive Office of the President. (1989b). *Special analyses: Budget of the United States government fiscal year 1990.* Washington, DC: Government Printing Office.

Villani, K. (1987). Finding the money to finance low-income housing. In P. D. Salins (Ed.), *Housing America's poor* (pp. 141-161). Chapel Hill: University of North Carolina Press.

Waldman, S. (1989, August 7). The HUD ripoff, *Newsweek*, pp. 16-22.

White, L. J. (1989). *The problems of the FSLIC: A policy maker's view.* Revised version of a paper presented at the annual meeting of the Western Economic Association, June.

Zigas, B. (1989). *The low income housing crisis and homelessness: The impact of federal policies 1981-1988.* Washington, DC: National Low Income Housing Coalition.

10

Mortgage-Backed Securities, Inflation-Adjusted Mortgages, and Real-Rate Funding: Recent Initiatives in Housing Finance in Australia

ROBERT A. CARTER

THE DEREGULATION OF FINANCIAL MARKETS

Substantial deregulation of financial markets occurred in Australia during the 1980s. This chapter explores the implications of deregulation for government involvement in housing markets—particularly that aimed at assisting marginal home buyers into home ownership. The central conclusion is that deregulation provided the opportunity for financial innovation in mortgage instruments and funding methods, and that this challenge has largely been taken up by the government sector.

At the commencement of the decade, numerous asset and interest rate controls were in place. In particular, savings bank home loan interest rates were regulated. Following the recommendations of the Campbell Inquiry (Committee of Inquiry into the Australian Financial System, 1981) and the Martin Report (Review Group into the Australian Financial System, 1983), substantial reforms were enacted. The major features of this liberalization included the entry of foreign banks, mergers between building societies that encourage greater diversification, the removal of many asset controls on the banking sector, and the deregulation of interest rates on savings bank home loans.

The removal of the protected status of the housing sector through deregulation potentially placed more home buyers on the margin of being able to afford home ownership by increasing home lending rates relative to other interest rates. In a society in which this is the majority (about 70%) tenure, a countervailing response became a political requirement. The commonwealth government's First Home Owner's Assistance Scheme provided means-tested relief for "deposit gap" problems in the form of tax-free grants. However, no mortgage interest deductibility has been provided.

The challenge was to exploit the positive aspects of deregulation through innovations that did not require major new subsidies to marginal home buyers. Developing the potential of the secondary mortgage market as a vehicle for real-rate funding and promoting the introduction of low-start, inflation-adjusted mortgages have been the most significant government-sponsored responses.

There has been considerable innovation in the private sector in the range and types of mortgage instruments and terms available; however, there has been very little evidence of a determined effort on the part of the major financial institutions aimed at increasing or maintaining the number of marginal Australian households who could obtain access to home ownership.

The lack of private activity aimed at the marginal home buyer market is explained by four major factors (Carter, 1987). First, the attractiveness of owner occupation (due to tax concessions and other incentives) assures a strong demand for finance from customers able to afford traditional (*crédit foncier*) mortgage instruments. Second, banks do not see the marginal home buyer as being their responsibility, and so conservative assessments of lending risk continue to prevail. Third, the switch from a traditional *crédit foncier* portfolio to low-start, inflation-adjusted mortgages on a significant scale could lead to short-term cash flow problems for the financial institution concerned. Finally, conservatism at branch manager level, combined with a perception of limited customer acceptance of different loan types, has discouraged widespread innovation in established financial institutions, particularly in the banking sector (Gloster, 1984, 1986).

Where the secondary mortgage market is concerned, private sector initiatives were initially directed toward the quality end of the market— particularly intérest-only loans in the investor market. Partly because of legislation, stamp duty, and other government-related restrictions, very little private sector initiative was directed toward generating affordable finance for the marginal first home buyer market.

The result has been that reforms intended to maintain the access of the marginal home buyer have largely depended on government initiative. Deregulation was the trigger for government-sponsored initiatives because it removed the traditional protections applying in housing markets. If deregulation was not to lead to a reduction in access caused by the removal of protection on interest rates, financial innovation was necessary. Deregulation also facilitated initiatives that previously were not feasible—particularly those related to funding from wholesale or secondary market sources.

This chapter then addresses the nature of mortgage innovation in a case-study government institution—the Ministry of Housing and Construction in Victoria. Following the outline of the inflation-adjusted mortgage approach, the chapter provides an overview of the government-sponsored development of mortgage-backed securities in Australia. The particular funding vehicle of Victorian housing bonds is introduced, and it is shown how this was a financial arrangement for the provision of capital-indexed Home Opportunity Loans. The chapter concludes with an assessment of the potential for further development.

THE USE OF INFLATION-ADJUSTED MORTGAGES

A major thrust of the financial innovations sponsored by the Ministry of Housing and Construction in Victoria has been the use of inflation-adjusted mortgage instruments. In the context of high and fluctuating nominal interest rates for housing, traditional *crédit foncier* loans make little sense for marginal home buyers. They result in a "front-end loading" problem in which the highest real payments on a mortgage occur in the initial years of the loan. This contrasts with the life-style pattern of many marginal home buyers, whose real incomes over time increase (associated with job promotion, additional family incomes, inheritance, and so on) rather than decline.

Through inflation-adjusted mortgages, the time path of repayments is shifted to one in which the real value of house repayments is steady over time. The purest form of inflation-adjusted mortgage is the capital-indexed loan. In the Australian context, its merits have been espoused by Stretton (1974), Carter (1986), and Sheehan and Derody (1982).

THE CAPITAL INDEXED LOAN PROGRAMME

Under a capital-indexed loan, a very low start in repayments results from the application of a real interest rate charge. Inflation is accounted for through adjustment of the balance outstanding (the nominal debt), rather than through alteration of the base interest rate. Experimentation with capital-indexed lending in Victoria commenced in July 1984 with the Capital Indexed Loan Programme (CAPIL).

The key features of the CAPIL loan are inflation-adjusted interest and repayments and a flexible loan term. The interest rate used is the actual rate of inflation plus a real interest rate. Commencing repayments are set at 25% of gross household income (an affordability consideration) and adjusted annually according to inflation in the past 12 months. Households can appeal on a needs basis if their incomes do not keep pace with inflation. The loan term is flexible because of the direct link between income and repayment levels.

Shifting to a mortgage instrument of this kind required something of a revolution in the culture of home ownership. Not only has the Australian housing policy tradition been one of subsidized interest rates and controls on home lending rates, but there is also strong established consumer expectation that housing is different from other consumer commodities in that costs (repayments) do not rise with inflation over time.

The government-sponsored program for capital-indexed lending also involves counseling and consumer advice responsibilities in establishing schemes. This occurs through brochures and intensive counseling of applicants, backed up by a telephone inquiry system.

Lending on a capital-indexed basis involved a substantial consumer education program in which the culture had to change toward an understanding that home loan repayments would rise over time, similar to other items in the household budget. As most of the borrowers in the CAPIL program had previously been private renters, explaining this by reference to the expected regular increases in private rental payments proved a useful technique.

The other major issue in consumer education with the capital-indexed loan is the notion of steadily increasing real equity in property, rather than the myth of a sudden rise to the status of full home owner. With a negatively amortized mortgage that has rising nominal debt for some years, the concept of increasing real equity has to be communicated to borrowers. The capital-indexed loan instrument sees the principal retired on an orderly basis over the life of the loan.

Consumer acceptance is only one of the difficulties in the transition to capital-indexed lending; there are also risks perceived in lending to low-income households through a negatively amortized mortgage. The two categories of risk often identified are that incomes will not rise in line with inflation and that property values in particular submarkets may fail to keep pace with inflation.

The response to these criticisms must be that the objective of these loan instruments is to tailor mortgage repayments better to the pattern of household life-cycle incomes. It is, therefore, a matter of lending to the right group of customers. Recognizing that the capital-indexed loan repayments schedule (in real and nominal terms) is identical to that of the *crédit foncier* system under circumstances of zero inflation places these criticisms in perspective, as being based on "money illusion." As Stroud (1981) points out, "The adoption of indexed-principal loans does no more than to return lenders (and borrowers) to the same risk situation they faced in times of negligible inflation" (p. 253).

Regarding the property value risk, only the most pessimistic of assumptions about rates of growth in house prices exposes lenders to risk under capital-indexed loans. Even with a 95% loan-to-valuation ratio, house prices have to rise at only 75% of consumer price index (CPI) increases for house values to cover peak outstanding nominal debt. Historically, house prices have almost invariably grown at least as fast as the CPI.

EVALUATION OF CAPIL

Recognizing that concerns would nonetheless be expressed over the transition to capital-indexed mortgage instruments and that there were grounds for some concern over extending lending to very low-income groups, the then Ministry of Housing in Victoria commissioned a five-year evaluation of the first batch of CAPIL loans. The initial evaluation reports by the Australian Institute of Family Studies (1987, 1988) produced highly satisfactory results, including the following:

- Over half of the 481 CAPIL scheme participants surveyed experienced a significant reduction in direct housing costs in the move from private rental to home ownership. Former public tenants were paying slightly more than previously.
- The move into home ownership had not resulted in significant repair costs for the majority of the pilot group, at least in the early years. In fact, there is considerable evidence of nonessential renovations and "sweat equity" being invested.
- The majority of loan group families indicated that they had sufficient income to meet all expenses on time or were able to save: 72% of one-parent families indicated that they were meeting expenses and 17% indicated an ability to save. The equivalent proportions for two-parent families were 68% and 24%, respectively.

- After meeting mortgage repayments, rates, and insurance, the capital loan families still possessed a higher after-housing disposable income than the control group of families renting privately, and only a minority of families (14% in the second-round evaluation) reported difficulty in meeting monthly repayments.
- The average incomes of CAPIL families increased both in real terms and relative to average weekly earnings. As a result, CAPIL loan families were far more likely than renters in either the public or the private sector to state that they were better off now than a year ago (Australian Institute of Family Studies, 1988).
- Levels of satisfaction with a variety of housing characteristics were much higher for the loan group than for the private rental control group.
- No families had defaulted under the scheme, and the pattern of arrears was comparable to *crédit foncier* Ministry of Housing loan schemes, which were also going to a higher-income group.

Final conclusions regarding the success of the CAPIL scheme must await the conclusions of the five-year evaluation—particularly on the further experience with maintenance and rates expenses, the reasons for early exits from the scheme, and an analysis of arrears patterns. However, the findings of the first two reports are very encouraging in terms of consumer satisfaction and the ability of households to cope with the financial obligations of home ownership and the mortgage instrument.

DEVELOPMENT OF THE SECONDARY MORTGAGE MARKET

The success of the CAPIL scheme, in terms of consumer response, raised issues of the availability of adequate funds and the appropriate basis for funding. While successful joint ventures with private sector institutions had been arranged on partially indexed instruments, there was concern from that sector over fully indexed loans. This kind of instrument also created the potential for a greater mismatch of assets and liabilities using traditional funding methods.

Given the deregulation of financial markets that had occurred, the major area of potential for fund-raising outside the budget and the joint ventures arrangements was through the wholesale and secondary mortgage markets. In particular, the possibility of issuing mortgage-backed securities on a real-rate (indexed) basis opened up the prospects of a long-term, flexible funding source that would match the cash flows from capital-indexed mortgages. Inspiration was drawn from the success of a similar scheme of index-linked bonds and loans in Denmark (Ministry of Housing, Copenhagen, 1983, 1984).

The issue of mortgage-backed securities on an index-linked basis required an appropriate institutional framework. It was left to state governments to establish mortgage-market corporations individually.

The Victorian government established the National Mortgage Market Corporation (NMMC) in 1984. The New South Wales government followed soon after, with the First Australian National Mortgage Acceptance Corporation (FANMAC). The model for the establishment of the corporations involved an initial state government sponsorship, but with private sector control of the ongoing operations of the corporation. In Victoria, the state government has a 26% share holding in NMMC, with the remainder of the shares distributed among a range of banking and other financial institutions. Initially, both bodies concentrated their activities on the pooling of existing mortgages as security for relatively short-term, mortgage-backed securities.

Development of the activities of the mortgage market corporations has been steady, though not spectacular. They have undoubtedly made a contribution toward greater stability in the flow of funds for housing and have provided existing home lending institutions with increased flexibility in balancing the supply and demand for housing finance.

Although developing the pooling of mortgages was a significant step, the requirement for matching low-start mortgage instruments to long-term finance raised through mortgage-backed securities had not been met. This required the further step of developing models for the issue of pass-through securities or debt instruments backed by the creation of mortgages. This would then represent "new" fund-raising for housing that would generate particular kinds of (low-start) loans.

VICTORIAN HOUSING BONDS

In 1987, the then Ministry of Housing initiated the development of the Victorian Housing Bonds scheme for raising index-linked (real-rate) funds through mortgage-backed securities. The manager of the program was to be the National Mortgage Market Corporation.

The objective of the Victorian Housing Bonds program was to match cash flow from capital-indexed Home Opportunity Loans (HOLs) with appropriate funding instruments at least cost. Particular attention was to be paid to ensuring the availability of funds and to minimizing the interest rate risk involved with reinvestment of prepaid mortgages (Ministry of Housing and Construction, 1988). The fulfillment of this objective created a particular challenge in funding. Essentially, the notion was that of raising index-linked finance on a long-term basis to match the repayment profile on capital-indexed mortgages.

The funding solution was not as simple as that of developing a long-term indexed security to match the time period of capital-indexed loans. This was so for the following reasons:

- The objective of matching cash flows over the life cycle of the loans with the cash flow owing to bond holders was moderated by the constraint of achieving the cheapest possible cost of funds overall to the program. The market for index-linked securities was not well developed, and there was some cost premium on these kinds of securities. The response to this problem was to develop a flexible funding approach, which enabled access to a number of "market windows" on both domestic and overseas markets.

- The capital-indexed loan instrument's terms were not uniformly predictable, because clients on higher incomes who borrowed less would pay out a loan over a shorter period than would clients on lower incomes. These differences in period of repayment could be considerable (a range of 10 to 30 years). There was also the likelihood of some prepayment of loans.

To cover this category of risk, it was necessary to have some ingredient of short-term and fixed-rate securities. This would be determined through constant monitoring of prepayment patterns and the tracking of market opportunities and interest rate changes through the financial model.

After a formal signing of the structure in June 1988, a year later $258 million in Victorian housing bonds had been issued, with over three-fourths of this amount in the form of inflation-indexed bonds (which best matched the cash flow characteristics of the mortgages). There was also a small issue of capital-indexed bonds in the total, and a short-term revolving funds facility had been established.

HOME OPPORTUNITY LOANS

While Victorian Housing Bonds are the source of the funds, Home Opportunity Loans are the destination. These are capital-indexed loan instruments designed to assist households that cannot afford conventional loans from banks, building societies, and other private sector financial institutions. When those administering the scheme identify a person who can afford a conventional bank or building society loan, they direct the individual that way, because the objective is not to compete with private sector lending, but to help those on the margin of affordability.

HOLs are made available through a cooperative venture involving the Ministry of Housing and Construction, NMMC, and a network of loan originators and servicers (called *retailers*). The Ministry of Housing and Construction issues publicity brochures and receives application forms. It keeps the waiting list and performs the initial screening on the income and asset situations of applicants and provides them with a detailed self-assessment process to ensure their eligibility prior to attending an interview with a designated retailer. These retailers include major savings banks and cooperative housing societies.

The assessment of eligibility includes a decision on whether the applicant is eligible for a rate buy-down in the form of a 2% real interest rate concession in the first two years of the loan. Applicants who are eligible for the concession can devote a maximum of an initial 25% of gross household income to repayments. The nonconcessional group's loan size repayment capacities are assessed according to an initial 27% of gross income maximum repayment. For both groups repayments are indexed to the inflation rate from the initial repayment level.

Once the eligibility of an applicant is established and a decision is made on whether he or she is eligible for the 2% rate buy-down, loan settlement procedures are concluded, with the trustee working for the manager (NMMC). Repayments on loans then occur through a direct debit-payment system operating out of bank accounts. NMMC is responsible for ensuring that there are sufficient funds in the repayment accounts held by the trustee to pay bond holders as required. Should shortfalls arise in the program because cash flows are not matching perfectly, the indemnity provided by the state government through the Director of Housing and Construction is called upon by the manager. Where this is clearly only a short-run cash flow problem, a subordinated loan is made by the Director of Housing and Construction to the bond issuer. Where the shortfall is caused by losses, such as defaulting mortgages that fail to recover full costs or reinvestment losses, the Director of Housing and Construction pays an indemnity to the issuer in the structure.

Proper management of the fund-raising and loan management side of the program should mean, over the long term, a balance of inflows and outflows. However, cash flow support will probably be necessary during certain phases of operation.

The interest rate facing borrowers depends on the market conditions prevailing at the time of bond issue. They are guaranteed no change in the real interest rate on their loans for the first five years. Thereafter, they are adjusted at intervals according to general market movements, and following an assessment of the overall cost of funds in the Victorian Housing Bonds program.

Real interest rates were at a historically high level in Australia in 1988, and charges of 6.5% were made on the initial batch of loans. For those eligible for concessional loans, this effectively meant a real interest rate of about 4.5% for the first three years. Given these real interest rates and the capital-indexed nature of the mortgage instrument, it was possible to make loans without subsidy to households on incomes substantially below average weekly earnings in Australia.

With the concessional interest rate, it was possible to make loans to recipients of lower incomes, including those on social security—in particular, single-parent households and unemployment beneficiaries. The profile of households with substantially below-average earnings would be low-wage, single-income households and larger families in receipt of social security benefits.

FUTURE DEVELOPMENTS

The network of retailers involved in the scheme has expanded, enabling greater access and choice among potential borrowers. Diversification has also begun in the application of funds raised. The Ministry of Housing and Construction has worked with community housing groups in establishing a program of Common Equity Rental Co-operatives (CERCs). This third rental tenure is well known in Canada and Europe and has equivalents in the housing cooperatives in England. Equity in property is shared among a group of 7 to 20 households, rather than individual titles being held. A cooperative structure is used to manage administration, rent collection, and maintenance tasks of the cooperative. Purchasing of houses in this scheme commenced only in 1986, and only five CERCs were in existence at the beginning of 1988. However, significant expansion in the scheme began in 1989, facilitated by access to Victorian Housing Bond finance. This long-term, real-rate financing better matches the life-cycle income profile of cooperatives and their tenants. Indexed financing is also used to assist in the development of Danish rental cooperatives and in the Canadian Co-operatives Scheme (Bossons, 1985).

CONCLUDING REMARKS

Deregulation of financial markets in Australia provided both a challenge and an opportunity in terms of the objective of maintaining home ownership access. With interest rates no longer controlled, an answer to the problems created by high and widely fluctuating nominal interest rates on *crédit foncier* mortgages was needed. In addition, the opportunity for development of wholesale and secondary mortgage market funding vehicles was presented.

Private sector market diversification continued to target customers perceived to be low risk, and it was substantially left to government-sponsored programs to innovate in the financial instruments directed toward marginal home buyers. The risk factors and the consumer education challenges involved in moving toward capital-indexed lending instruments were absorbed by the Ministry of Housing and Construction in Victoria. The scheme introduced real-rate mortgages to the market and extended home ownership opportunities to a wider group without subsidy. It also created pressure for a higher level of funds to be made available on a basis that matched the cash flow from the mortgages. Victorian Housing Bonds (largely inflation indexed) were issued to provide a wholesale source of off-budget funds for the government-sponsored capital-indexed loan program.

In the context of deregulation and a general trend toward stringency in government expenditure, these programs are moving to center stage in terms of housing policy in Australia. In 1989 the capital-indexed loans scheme was extended to both Western Australia and South Australia.

The Australian experience also leads to the conclusion that deregulation does not alter market segmentation tendencies. There was no wave of innovation from the private sector in terms of either funding instruments or mortgages that would have broadened their customer base to include a lower-income group. Governments have become even more clearly responsible for their traditional market segment.

REFERENCES

Australian Institute of Family Studies. (1987, July). *First evaluation report of the Victorian Ministry of Housing's Capital Indexed Loan Pilot Scheme.* Melbourne: Author.

Australian Institute of Family Studies. (1988). *Home ownership for low income families: Second evaluation report of the Victorian Ministry of Housing's Capital Indexed Loan (CAPIL) Scheme.* Melbourne: Author.

Bossons, J. (1985). *The effects of using index mortgage-backed securities to finance non-profit housing co-operatives.* Ottawa: Co-operative Housing Foundation of Canada.

Carter, R. A. (1985, February). Emerging housing policy issues in Australia: Deregulation, tenure neutrality and all that. *Australian Urban Studies,* pp. 8-14.

Carter, R. A. (1986). *Changes in the housing policy context in Australia in the 1980s: The role of restructured housing mortgages and housing finance innovation.* Paper presented at the International Research Conference on Housing Policy, Gavle, Sweden.

Carter, R. A. (1987, February). Innovations in financial mechanisms for housing markets: Deregulation and fiscal conservatism. *National Economic Review, 6.*

Committee of Inquiry into the Australian Financial System (Campbell Inquiry). (1981). *Final report.* Canberra: A.G.P.S.

Gloster, G. (1984, July). Explaining the myths surrounding housing finance. *Housing Australia,* pp. 12-16.

Gloster, G. (1986). *The potential for more relevant housing finance packages.* Paper presented at the Seminar on Finance for Private Rental Housing, Melbourne.

Halifax Building Society. (1986). *The Halifax matched index-linked finance portfolio* (Background Information Paper). Halifax, England: Author.

Ministry of Housing, Copenhagen. (1983). *Index-linked financing in Denmark.* Copenhagen: Author.

Ministry of Housing, Copenhagen. (1984). *Financing of housing in Denmark* (rev. ed.). Copenhagen: Author.

Ministry of Housing, Victoria. (1985). *New mortgage instruments for the 1980's: Victorian government initiatives with inflation-adjusted mortgages.* Melbourne: Author.

Ministry of Housing and Construction, Victoria (1988). *Mandates under the Victorian Housing Bonds Programme* (Internal Paper). Melbourne: Author.

Modigliani, F., & Lessard, D. (Eds.). (1975). *New mortgage designs for stable housing in an inflationary environment.* Boston: Federal Reserve Bank of Boston.

National Mortgage Market Corporation Ltd. (1987). *Annual report, 1987.* Melbourne: Author.

Priorities Review Staff. (1975). *Report on housing.* Canberra: A.G.P.S.

Review Group into the Australian Financial System (Martin Report). (1983). *Report of the Review Group into the Australian Financial System.* Canberra: A.G.P.S.

Sheehan, P., & Derody, B. (1982). The Campbell Report: A critical analysis. *Australian Economic Review,* 1st quarter, 35-58.

Spiller, M. L., & Carter, R. A. (1986). *Finance deregulation: Intrametropolitan impacts and the future direction of federal housing policy.* Paper presented at the World Planning and Housing Congress, Adelaide.

Stretton, H. (1974). *Housing and government* (1974 Boyer Lectures). Sydney: Australian Broadcasting Commission.

Stroud, J. F. (1981). *Housing finance interest rate ceilings: A comment on the possible consequences of deregulation* (Australian Financial Systems Inquiry, Commissioned Studies, Part 4). Canberra: A.G.P.S.

Supervisor of Mortgage Finance for Denmark. (1985). *Annual report.* Copenhagen: Author.

11

Deregulation of Private Rented Housing in Britain: Investors' Responses to Government Housing Policy

ANTHONY D. H. CROOK

THE BRITISH GOVERNMENT ENACTED legislation in 1988 designed to reverse a long-term decline in private renting. New lettings have been deregulated and tax breaks given to individuals investing in private renting. This chapter discusses the likely impact of these measures. First, housing policy objectives in the 1980s and contemporary private rented housing and households are described. Second, recent investment is examined, showing how it was shaped by current policy. Third, the needs of tenants and landlords that policy should meet are established. Fourth, the deregulation policy and its demand and supply effects are discussed.

HOUSING POLICY IN THE 1980s AND THE PRIVATE RENTED SECTOR

HOUSING POLICY OF THE 1980s CONSERVATIVE GOVERNMENTS

Until the 1980s, postwar governments saw housing as having significant social value and as a suitable vehicle for redistributive policies. Intervention was accepted, both in direct (though varying) provision of social rented housing and in the regulation and promotion of private housing. The Thatcher government, by contrast, believes housing

AUTHOR'S NOTE: The research on which this chapter is based was funded by the Joseph Rowntree Memorial Trust, whose assistance is gratefully acknowledged.

should be privatized. Above a basic minimum, housing is a private good, benefiting individuals, not society as a whole. The state's job is to define the minimum and provide income assistance where the poor cannot afford it. The market is, however, more efficient in providing housing and meets preferences at less cost in resources than public provision and bureaucratic allocation (Whitehead, 1983).

The privatization argument is ideological as well as economic. Increasing owner occupation per se, not just private ownership, is important because of the benefits the government believes society reaps from the attitudes and commitments of owner-occupiers to "social stability." It has therefore maintained and extended subsidies to owner-occupiers.

Between 1979 and 1987, conventionally defined public expenditure on housing was cut by 60% in real terms, resulting in big falls in new social rented provision and steep real rises in social housing rents. Private investment was substituted for public spending, including specific low-cost home ownership projects and a more competitive mortgage market. Social rented stock was transferred to owner occupation through tenants' "right to buy" at discounts.

If the first two Thatcher governments relentlessly pursued home ownership, the third one (from 1987) is giving greater attention to rented housing, although the promotion of owner occupation is to continue (Department of the Environment, 1987). Local authority direct provision is being ended. In its place, commercial private renting is to be revived, nonprofit housing association rented provision will grow, with mixed private and public funding, and existing local authority estates are to be transferred to a range of nonstatutory landlords.

These changes reflect the growing interest by government and independent commentators in Britain and abroad in reviving private investment in rented housing (Kemp, 1988; National Federation of Housing Associations, 1985). The reasons include a growing concern about the shortage of rental housing in a period of fiscal austerity, the way the physical decay of private rented stock undermines inner-city renewal programs, the need to house workers moving from economically distressed peripheral regions to jobs in areas of economic growth, and, in Britain, the lack of choice for anyone wanting to rent tolerable housing in desirable neighborhoods and the government's preference for privatization and hostility to direct provision by elected local authorities (many controlled by the Labour party). Independent commentators recognize, however, that reviving private renting in Britain requires a restructuring of housing subsidies to enable private renting to compete with other tenures.

A PROFILE OF
PRIVATE RENTING IN BRITAIN

By 1987, only 8% of all British households rented privately: 2% from employers and 4% as long-term unfurnished lettings, leaving only 2% let as ready-access furnished housing. Two-thirds of rental housing was built before 1919, and physical conditions in rental housing are poorer than in other tenures. In 1986, 35% needed more than £1,000 in repairs, compared with 12% of owner-occupied stock (Department of the Environment, 1988). The very worst conditions are found in houses in multiple occupation, 80% being below amenity, management, and occupancy standards in 1985 (Crook, 1989; Thomas with Hedges, 1986).

Private renting has an increasing overrepresentation of nonfamily households outside the labor market (Bovaird, Harloe, & Whitehead, 1985; Todd, Bone, & Noble, 1982; Todd & Foxon, 1987; Whitehead & Kleinman, 1986). The unfurnished subsector houses mainly elderly households and others of long-standing residence, providing secure and low-cost, but low-quality, housing. Landlords tend to sell rather than relet vacancies. The rapid-turnover furnished subsector houses young singles and other mobile nonfamily households, meeting the demand for short-stay, ready-access housing with low transaction costs, but in poor condition, insecure, and at high cost. Only a minority of the young are actually in private renting, however (only 15% of all those under 30 years old in 1985). Most of this accommodation is relet upon vacancy. Private renting also houses those in tied, job-related accommodation and provides housing for those who would prefer but cannot achieve other tenures.

Private renters have low incomes. Median household incomes of unfurnished and furnished private renters were 48% and 80%, respectively, of the median for all tenures in 1985, reflecting the numbers who are economically inactive and the low pay of those in the labor market. Both social and private rented tenants have become increasingly concentrated at the bottom of the income spectrum over the past 30 years (Bentham, 1986). Social landlords, however, have had subsidies on their stock from central government, while private landlords have not. Without subsidies a considerable gap has opened up between the rents private landlords need to get competitive returns and what low-income tenants can afford.

THE DECLINE OF PRIVATE RENTING
AND THE LEGAL FRAMEWORK

Levels of demand, political uncertainty, poor reputation, and tax/ subsidy policy have been just as important as controls in explaining decline (Doling & Davies, 1984; Hamnett & Randolph, 1988; Harloe, 1985; Holmans, 1987; House of Commons Environment Committee, 1982; Nevitt, 1966). Decline has been accompanied by the expansion of owner occupation, aided by tax concessions on mortgage interest and fueled by inflation, so that landlords cannot offer housing at rents competitive with home purchase costs for those who can afford to buy, while subsidized social rented housing provides alternatives for those who cannot buy. Middle- and upper-income groups who used to sustain private renting now own their own homes. Demand is limited to young singles and the elderly with meager incomes to pay the rents landlords need for competitive returns. It is this low demand, unaccompanied by supply subsidies, that is the cause of decline, not rent and security regulation. Previous attempts to deregulate have not led to a revival, but rather to abuse by landlords like Rachman, notorious in London in the late 1950s (Committee on Housing in Greater London, 1965), contributing to the poor image of landlords. "Rachmanism" has haunted private renting and threatened to undermine attempts to get "responsible" investment.

By 1979, the Rents Acts gave most private tenants statutory security of tenure. Tenants could not be evicted, except in statutorily defined circumstances and only with the courts' consent (Arden, 1989). In certain circumstances relatives can succeed to tenancies. Most tenants in the unfurnished long-term subsector were protected, but many lettings in the rapid-turnover subsector were put beyond the scope of protection by landlords who deliberately used devices to evade the legislation, like letting on nonexclusive occupation agreements and the sham use of genuine Rent Act exclusions, such as holiday lets. Up to half of recent new lettings in the 1980s were made in this way (Greater London Council, 1986; House of Commons Environment Committee, 1982; Todd & Foxon, 1987).

By 1979, tenants and landlords could agree upon rents themselves, but could alternatively get a rent officer (a public official) to register a Fair Rent--a valuation reflecting all the circumstances of a tenancy, but excluding any excess demand arising from shortages. This becomes the maximum rent, subject to periodic review. Most long-term unfurnished rents are Fair Rents. Most short-term rapid-turnover tenancies' rents are privately agreed upon. In addition, licensees, and others excluded from Rent Act protection, cannot get Fair Rents registered. The Fair Rent system was set up in 1965 to replace progressively the earlier practice of rent control (rents fixed by statute) and was designed to move toward market rents while protecting tenants from the impact of shortages. Rent allowances to help pay these higher rents are part of the housing benefit system administered by local authorities, whose costs in paying benefit are subsidized by central government (Kemp, 1987b). Allowances are means tested, unlike the tax reliefs of owner-occupiers with mortgages.

Standards in respect to amenities, occupancy, and repair need to be set and monitored, since landlords may attempt to reduce them if regulation restricts their revenue. Local authorities can insist on improvement and repairs, and landlords (like all owners of older private houses) can get grants for some of these (Crook, 1989; Crook with Sharp, 1989).

PRIVATE RENTED POLICY
IN THE 1980s

Up to 1980, the balance struck by policy between the interests of existing and potential tenants and landlords had broadly favored existing tenants. In 1980 the government moved the balance toward landlords and potential tenants. The Housing Act 1980 introduced two new forms of tenancy that gave landlords two alternatives to letting within the existing Rent Act framework described above. First were shorthold tenancies, which guaranteed landlords who let vacant houses repossession at the end of a term of one to five years. Second were assured tenancies, which allowed government-approved companies to let vacant dwellings at market rents with contractual security. These initiatives had little impact because they failed to tackle the underlying questions of demand and subsidy discrimination (Crook, 1986; House of Commons Environment Committee, 1982; Kemp, 1987a).

In 1987 the government announced it would go much further and deregulate all new tenancies. It argued that private renting was a good option for mobile people who did not want the ties of home ownership, that rent regulation reduced landlords' returns, giving them no incentive to stay in the market, and that statutory security deterred people from letting temporarily. Partial deregulation was proposed to stimulate supply (Department of the Environment, 1987).

INVESTORS' RESPONSES TO
PREDEREGULATION POLICIES IN THE 1980s

Two sources of evidence are used to see how policies shaped investment up to 1988, particularly who invested and why. First is a 1985-86 follow-up survey of a panel of private rented addresses and owners in inner Sheffield (a major industrial city in northern England) originally set up in 1979-80. The panel comprised 1,377 addresses owned by 845 private landlords in 1979-80 and was a representative sample of the private rented sector in that city (Crook & Bryant, 1982; Crook & Martin, 1988). Second is a 1987 survey of 41 urban local authorities, representative of urban areas in the North and Midlands regions, and having 77% of all private renting households in these regions (Crook, 1989; Crook with Sharp, 1989). This survey confirmed the general applicability of the Sheffield findings with respect to recent investment. Evidence from official surveys of landlords and evidence of a parliamentary select committee confirmed that the evidence found in this study about landlord behavior in the North and the Midlands was also to be found in other regions (House of Commons Environment Committee, 1982; Paley, 1976; Todd & Foxon, 1987).

INVESTMENT AND INVESTMENT RETURNS
IN THE LONG-TERM UNFURNISHED SUBSECTOR

The follow-up survey in Sheffield revealed that, while the unfurnished subsector had declined by nearly one-third in the six-year period (as landlords sold vacant properties to owner-occupiers), there was also significant new investment going into the subsector. This did not involve the purchase of vacant properties or the building of new ones for letting. Instead, long-standing individual landlords were selling off property with sitting tenants to large company landlords, acting like property dealers, often in the building industry and new to landlordism. Some 20% of the 1985 Sheffield panel changed hands this way in the previous six-year period.

This was not long-term investment. In Sheffield, and in two-thirds of the 41 urban authorities studied, property dealers were speculating by buying properties with sitting tenants to sell when these tenants left. In the meantime, properties were improved with grants from the local authority, making them salable and mortgageable and providing building work for the landlords' businesses. They bought run-down property (to attract grants), with single pensioner sitting tenants (to maximize chances of vacancy and avoid relatives with succession rights), and on a large scale (guaranteeing a steady rate of vacancies). Evidence presented to the parliamentary select committee confirmed that this speculative investment was the only area of significant investment in private rented unfurnished housing (House of Commons Environment Committee, 1982).

Between 1979 and 1985, the rents of the Sheffield panel doubled in real terms; 80% of these were registered Fair Rents, a proportion reflecting the national level of increases in Fair Rents. Rates of return from rents were still low. Returns from unfurnished properties in 1985 (based upon rents and the 1985 vacant possession capital values) were only 5% per annum gross (3% net of management and maintenance costs) on rents alone (i.e., making no allowance for capital appreciation).

Only 32% of unfurnished houses had landlords who were satisfied with these returns, making comparisons with returns from more liquid and less risky alternatives like building societies or government bonds. When real increases in rents *and* capital appreciation are also taken into account, landlords who bought unfurnished sitting tenant property in 1979 are shown to have received more returns than by placing the sitting tenant price in either building societies or equities over the period 1979 to 1985. Landlords were, nonetheless, rational to say they would sell unfurnished property if it became vacant. If the Sheffield landlords had acquired vacant possessions in 1979, for example, they would have been better off investing the sale proceeds in alternative investments than they would have been to continue renting. *Only* if they could have been certain in 1979 that they would get vacant possession (and capital gain) in 1985 would they have been better off staying in property. Thus, despite the doubling of rents in real terms since 1979, 80% of Sheffield panel properties had landlords who would sell vacancies "tomorrow." This finding is consistent with other survey evidence (for example, see Paley, 1976).

If landlords can expect real increases in rents and capital gain, it may be better to compare nominal returns from renting with equities, rather than with building societies or bonds. Nevertheless, the 3% nominal net return was clearly inadequate, given the low liquidity and high risk of private rental investments at that time. The British Property Federation in 1982 argued that 6% net, 9% gross was needed, implying a doubling of rents on existing older stock in Sheffield and a quadrupling on new development, along with adjustments to security of tenure to increase liquidity (House of Commons Environment Committee, 1982).

INVESTMENT AND INVESTMENT RETURNS
IN THE RAPID-TURNOVER
FURNISHED SUBSECTOR

This subsector also had new investment. Over a quarter of the 1985 Sheffield total had changed hands since 1977, some, significantly, being bought with vacant possession by new owners. Most landlords were small-scale, with fewer than six properties in all. Almost all the 41 authorities studied in 1987 experienced an increase in furnished accommodation.

In Sheffield landlords bought run-down houses for rent income from continued letting, rather than for capital gain, and "milked" them for the maximum net rental. Furnished property was in worse repair than unfurnished in both 1979 and 1985. New landlords acquired furnished property in the very worst repair, and few had put any defects right by 1985.

Tenants did not complain. They neither understood the system nor felt confident about complaining, whether to local authorities or to landlords. They feared landlords would evict them as a result and they knew there was a shortage of good-quality affordable housing. This acceptance of poor housing is confirmed by other research (Thomas with Hedges, 1986). The tenants' insecurity is highlighted by the fact that most Sheffield lettings were unprotected, half being let outside the Rent Acts by their landlords to restrict their tenants' security and prevent Fair Rents being registered, a finding consistent with national survey evidence (Todd & Foxon, 1987).

Rents of furnished houses in Sheffield rose 25% more than inflation between 1979 and 1985. Rents were de facto already deregulated in 1985, only 7% of rents being registered. High proportions of gross incomes went in rent, a finding again consistent with evidence from national samples (Todd et al., 1982).

Average annual nominal rates of return were 14% gross and 9% net in 1985. Not surprisingly, 72% of furnished houses in Sheffield had landlords satisfied with their rental income, and 80% of furnished houses would have been relet in 1985 if they had become vacant "tomorrow."

PROBLEMS FOR TENANTS AND INVESTORS: IMPLICATIONS FOR POLICY

In the unfurnished long-term subsector the Sheffield evidence showed that property dealers had made money from letting property in good condition to low-income people, but only by investing for short-term capital gain rather than for rent. Rent increases after 1979 were not enough to persuade them to stay in residential property by closing the gap between sitting tenant values (the price investors pay to buy the discounted stream of net rents and capital appreciation) and vacant possession values (the price owner-occupiers are willing to pay). This is because rents were kept low by regulation and by the low effective demand of poor tenants, because of the influence of tax concessions on the prices owner-occupiers will pay for vacant houses, and because tenants' security puts restraints on landlords realizing this price.

Rents did not give landlords returns on vacant possession value commensurate with returns elsewhere, with similar risk and liquidity. Existing landlords therefore sold up when they could. The value gap also explained the activities of property dealers who exploited it by buying at sitting tenant value to sell at vacant possession value. Without substantial extra subsidies, rents in a deregulated market paid by low-income tenants are unlikely to rise sufficiently to close this value gap and so prevent continued decline of the existing stock, let alone rise to levels that would give competitive returns on new building.

Although nominal rates of return in the rapid-turnover furnished sector were comparable with alternative liquid investments, the Sheffield survey and other evidence suggests that profitable letting to low-income people was possible only by renting dilapidated and badly managed older housing let outside the Rent Acts at high occupancy rates to tenants paying very high proportions of their gross income in rent. The need to treat tenants in this way in order to make profits reinforced negative images of landlordism.

While this subsector was de facto deregulated by 1985, tenants were insecure and got poor value for their money. For landlords, demand depended on the income maintenance systems on which tenants depended, costs depended on the local authorities' willingness to enforce repair and other standards actively, and they had to let outside the Rent Acts. Until 1988 this was a high-risk area, and a risk premium was required to make returns competitive. As a result, the pre-1988 framework met the needs of neither landlords nor tenants. Most unfurnished tenants paid (relatively) low rents for secure but poor-quality properties. Furnished tenants paid high rents for insecure and poor conditions. Prospective tenants searched under conditions of excess demand. Landlords either accepted below-market returns or operated outside the legal framework. The dilemma is that rents would have to rise considerably to give landlords competitive returns on decent housing, but such rents could not be paid without hardship or considerable subsidy (House of

Commons Environment Committee, 1982). Deregulation per se is no guarantee that market rents would rise to levels giving competitive returns. If these problems are to be resolved within a market framework, policy must address questions of demand as well as supply.

Tenants in the long-term subsector are mainly elderly and require low rents, security, and improved standards. They need protection from harassment if landlords seek to sell or to let in the deregulated sector. Ultimately, landlords will sell upon vacant possession. Meanwhile, they need incentives to improve and maintain standards, but there is no case for deregulating rents, since most landlords have paid a sitting tenant price and are getting adequate returns on this.

Tenants in the rapid-turnover sector want easy access, habitable standards, and affordable rents more than long-term security. Few can pay rents that give them these things while also giving landlords adequate returns. Landlords in the rapid-turnover sector want adequate returns in relation to the liquidity and risk of this type of investment. That means a combination of rents that gives competitive returns, the ability to evict bad tenants, and the right to repossession when they want to sell. Stability in the legal framework is needed, to eliminate uncertainty about the reintroduction of controls. It is also important to increase reputation. If landlordism is seen as a repugnant form of investment, many will not enter the field, even if the returns are good. Capital subsidies may be important in ensuring both that returns are competitive and that social objectives are achieved.

DEREGULATION

THE GOVERNMENT'S MEASURES

· The relevant provision of the Housing Act 1988 came into force in January 1989. Existing tenants lost neither their security nor their right to Fair Rents. Laws on harassment were strengthened, and tenants can get damages, if illegally evicted, up to the amount of profits landlords make from selling the vacant property.

All new lettings (including relets), however, are at market rents, but with varying security. There are two ways of letting. First, there is a modified form of assured tenancy with freely negotiated rents, where security will be protected by fixed tenancies running on as statutory periodic tenancies, subject to new mandatory and discretionary grounds for courts to sanction eviction, such as, respectively, three months' rent arrears and persistent delays in paying. If the basis for rent increments is not specified in the contract, a Rent Assessment Committee may adjudicate on a market rent. The previous requirements for assured-tenancy landlords to be approved and for qualifying lettings to be newly built or improved has been abolished. Second, there are assured shorthold tenancies, which will have no security beyond a fixed period (as little as six months) but tenants may in certain circumstances apply to the Rent Assessment Committee to determine a market rent that will take account of the limited security.

Market rents paid by housing benefit claimants are referred to rent officers for validation as reasonable to prevent the subsidy the government provides on local authorities' housing benefit payments being paid on excessive rents. Guidance on the method of calculating reasonable market rents will be based on a rate of return method allowing for capital appreciation.

The government also intends to restructure the grant system for improving older housing, and legislation has been incorporated in a bill that is before Parliament at the time of this writing. Standards are to be reduced. All grants to private landlords are to be discretionary. Grants paid will be related to a test of landlords' ability to finance the eligible work out of rental income rather than the current system of fixed percentages of eligible costs.

In conjunction with the deregulation measures, tax incentives have been given to invest in private rented housing. For the first time since the 1920s, they provide subsidies for the provision of private rented housing, enabling private landlords to compete on more nearly equal terms with other tenures. Private investors have been given both income and capital gains tax incentives to invest in new residential property companies set up to let new assured tenancies. Apart from maximum limits placed on the price that companies can pay for properties, there are no other targets restricting the application of funds in terms of qualifying properties or tenants, unlike, for example, the low-income housing tax credit in the United States (Gruen, 1989).

EVALUATION:
THE CONTINUING REGULATED SECTOR

The government has been correct not to remove controls from existing tenants and to tighten up the laws on harassment. Ultimately, the properties will be sold off, mainly to owner-occupiers. How far the grant reforms will remove property dealers' and long-standing owners' incentives to carry out improvements before then depends on the test used to define the rental resources available to pay for improvement, whether it is property or firm related, how much is allowed for risk and liquidity, and how much expenditure is to be eligible for grant aid.

If incentives are reduced, failure to get property improved will be costly, not only to many tenants, but to whole neighborhoods. The government needs to tackle this problem on a much more comprehensive basis, combining mandatory and discretionary grants with tax allowances to encourage investment in the stock. There is a strong case (provided that tenants' interests are protected) for encouraging a private sector "buy-out" from older long-standing landlords who are uninterested in improvements (Maclennan, 1986).

EVALUATION:
THE DEREGULATED SECTOR

The government's aims are to increase rents, to reduce risk, and to enhance liquidity, thereby increasing returns to a competitive level and drawing in new, responsible investors who will expand the overall supply of rental housing for persons at early stages in their life cycles and for job movers. It does not intend private renting to provide for families or other low-income households. This is correct; owner occupation or social rented housing can better meet the needs of families and the vulnerable. Only those with economic power are likely to benefit from the attributes of private rented housing in relation to ready access, lower transaction costs, fewer obligations, and flexibility in changing housing decisions (Whitehead & Kleinman, 1986). Given the existence of mortgage-interest income-tax relief for owner-occupiers, renting at market rates without an equivalent fiscal or other subsidy for tenant or landlord is an attractive option only in limited circumstances, such as for people who move often and incur a lot of transaction costs, those who need to save up for a deposit to buy, or short-term residents, like students and overseas businesspeople (Coleman, 1989).

The market for this ready-access housing was, however, de facto already deregulated before 1988. The question is whether de jure deregulation will lead to increased rents, provide competitive returns, draw in new investors, increase supply, and give tenants bargaining power.

On the demand side, there are at least three reasons to suggest that it will be difficult for low-income tenants to pay higher rents. First, students (who are a significant element of potential demand) have had grants cut in real terms by 20% since 1980 and have lost some of their eligibility to claim housing benefits. Student grants are to be frozen and topped up with loans after 1990, and this increases uncertainty about levels of demand from students.

Second, the revised housing benefit system introduced in April 1988 worsened the position for those in work, those with occupational pensions and savings, and those under 25 (Kemp, 1987b). Significant numbers, including low-paid workers, receive little help toward total rent, and those on benefit are subject to high marginal tax rates if their incomes increase, because of the steep taper of 65 pence in every pound by which benefit is withdrawn when net income exceeds income support, the amount at which their housing costs are met in full.

Because all tenants in receipt of some housing benefit receive, in extra benefit, the full amount of any rent increase, they have marginal housing costs of zero and no financial incentive to "haggle" with landlords over rents and to "shop around" to find alternatives. As a result, the rents of claimants are to be referred to rent officers to check whether the rent is a market rent or one that reflects the operation of the housing benefit system. If these validated rents are set below actual rents, local authorities will be free to pay full benefit to tenants on the rent actually charged, but will have to pay for the full cost of the difference and will not receive reimbursement on it from central government. If they prefer to pay benefits only on the validated rent, tenants will then have to pay even more from their meager resources, or landlords will increase occupancy rates for a lower per person rent, or rents will fall, causing landlords to leave the market.

Thus, while the formal removal of controls may lift any ceiling currently placed on de facto deregulated rents, there is little evidence that the market can sustain substantially higher rents, given the low income of tenants and the new housing benefit system. Most tenants will be too poor to pay the rents that will give landlords competitive returns, while those who can pay them will find it cheaper to become home owners. If rents rise, it will give further incentive to people who are ineligible for housing benefits to transfer into owner occupation, where they will get tax relief on mortgage interest.

That is not to say, of course, that a growth in supply of new or well-improved privately rented houses in desirable neighborhoods would not, of itself, create a demand that does not now exist. Given that such accommodation is very difficult to find now, potential demand is deflected into owner occupation, where such property is more readily available. Potential demand for such rented accommodation may well be found among groups of job movers and other well-paid professionals and other workers at early stages of their careers who value the flexibility that renting can bring and who can also afford rents that give landlords competitive returns. Much of the recent investment in new property companies triggered by the new tax incentives is targeted at such groups. Currently many such tenants choose to buy and tie up capital, not only because subsidized home ownership is very competitive compared with paying an economic rent, but because they cannot find housing of comparable quality in the rented sector.

The new legislation also provides some of the necessary conditions for increased supply. In principle, the legislation enhances the reputation and increases confidence by lowering the risk of this class of investment, since it provides a statutory framework for letting at market rents with minimal security and reduces the risks of being saddled with "bad" tenants. The rate of return required for this lower risk should be less than under the old framework. Given market rents, contractual arrangements for rent reviews in line with inflation and statutory powers to remove bad payers, together with other possession rights enhancing liquidity, some of the requirements of larger, long-term investors could be met within the new framework. In principle, therefore, the legislation should overcome many of the problems landlords experienced under the old framework and should meet many of their needs.

In practice, it is simply not clear that deregulation will result in rents that give competitive returns without significant subsidy. The earlier assured-tenancy scheme foundered after the withdrawal of capital allowances, because investors considered returns to be inadequate from rents at current levels of demand (Kemp, 1987a). The new tax incentives are an attempt to reintroduce capital subsidy for a short period, although it remains to be seen whether it will create only a short-lived "wave" rather than more enduring investments. The first signs from prospectus are that it will be short-lived, most promoters emphasizing capital growth and "exit routes" for investors to achieve tax-free gains. Even more important, from the point of view of bringing in long-term responsible investors, is that there is no political consensus and no certainty about the long-term stability of the legal framework. The legislation will have to survive at least one change in government before confidence returns.

Reputation--which is crucial to institutions--could be undermined by the government's determination not to resurrect the system of prior approval of assured-tenancy landlords. Instead, the government prefers the approach of policing standards, and has strengthened local housing authorities' repair enforcement powers.

There may, however, be some expansion among smaller individual landlords who get the power via assured shorthold tenancies to realize vacant possession values and remove bad tenants easily. Such landlords might be prepared to accept lower rents and rates of return than the investors described above and may provide better value for money and improved standards, especially for young people.

CONCLUSIONS

It is hard not to conclude that the 1988 legislation fails to address the underlying problem of the private rented sector concerning levels of demand, competing tax and subsidy arrangements in other tenures, and confidence. Capital subsidies may be the only obvious way of ensuring both that returns are competitive and that social objectives are achieved. Given all the risks of providing low-rent housing to low-income people and the uncertainties about the rent allowances, it is more likely that investors will respond to capital subsidies than make investments whose returns are dependent on tenants' rent-paying capacity being underwritten by housing benefits. At the same time, if other tenants on higher incomes are to be attracted, rents need to be set at levels that makes renting competitive with house purchase. Again, capital subsidies may be needed. Without the abolition or phasing out of mortgage-interest relief, some compensating subsidy to private renting is required, and the new tax incentives are a start. These should be allied with greater use by local authorities of revised discretionary and mandatory powers to police and enforce standards of houses let at market rents. A system of prior approval of landlords is also needed (which would increase the likelihood of bipartisan support) and retention of approved status should be subject to letting at acceptable minimum standards.

On the demand side, tenants need a greater degree of market power if the proposals are to work to their advantage. First, prescribed forms of contracts should be required for all new lettings, together with the establishment of housing courts to arbitrate on disputes. Second, the housing benefit system needs to be remodeled fundamentally. Consideration should be given to basing rents on standard market rents valued for each area. Claimants would be allowed to keep any difference between the standard and actual rent agreed upon with the landlords. At the risk that some tenants would lose, the gain would be giving tenants both incentive and power with which to bargain. Third, tapers on benefits for those whose net income exceeds income support levels should be less steep, and allowances used to calculate eligibility should be more generous. Fourth, the bargaining power of low-income tenants would be greatly increased if their access to social rented housing were increased.

All of this means both more public investment in the provision of adequate rented housing and more public expenditure on helping low-income tenants afford their standards. Unfortunately, this conflicts with government policy to cut public spending on housing and to maintain fiscal support for owner occupation, which reduces demand for private renting. Revival is unlikely on these terms.

REFERENCES

Arden, A. (1989). *Manual of housing law* (4th ed.). London: Sweet & Maxwell.

Bentham, G. (1986). Socio tenurial polarisation in the UK 1953-1963: The income evidence. *Urban Studies, 23*, 157-163.

Bovaird, A., Harloe, M., & Whitehead, C. M. E. (1985). Private rented housing: Its current role. *Journal of Social Policy, 14*, 1-23.

Coleman, D. A. (1989). The new housing policy: A critique. *Housing Studies, 4*, 44-57.

Committee on Housing in Greater London. (1965). *Report* (Cmnd 4609). London: HMSO.

Crook, A. D. H. (1986). Privatisation of housing and the impact of the Conservative government's initiatives on low cost home ownership and private renting between 1979 and 1984 in England and Wales: 4. Private renting. *Environment and Planning A, 18*, 1029-1037.

Crook, A. D. H. (1989). Multioccupied housing standards: The assessment of discretionary powers by local authorities. *Policy and Politics, 17*, 41-58.

Crook, A. D. H., & Bryant, C. L. (1982). *Local authorities and private landlords: A case study.* Sheffield: Sheffield Centre for Environmental Research.

Crook, A. D. H., & Martin, G. J. (1988). Property speculation, local authority policy and the decline of private rented housing in the 1980s. In P. Kemp (Ed.), *The private provision of rented housing.* Aldershot: Avebury.

Crook, A. D. H., with Sharp, C. B. (1989). *Property dealers, the private rented sector and local authority policy in urban areas of the North and Midlands* (Occasional Paper). Sheffield: University of Sheffield, Department of Town and Regional Planning.

Department of the Environment. (1987). *Housing: The government's proposals* (Cmnd 214). London: HMSO.

Department of the Environment. (1988). *English house condition survey 1986.* London: HMSO.

Doling, J., & Davies, M. (1984). *The public control of private rented housing.* Aldershot: Gower.

Greater London Council. (1986). *Private tenants in London.* London: Author.

Gruen, N. J. (1989). *The low income housing tax credit and private sector investment in new and rehabilitated affordable housing.* San Francisco: Gruen, Gruen and Associates.

Hamnett, C., & Randolph, B. (1988). *Cities, housing and profits: Flat break up and the decline of private renting.* London: Hutchinson.

Harloe, M. (1985). *Private rented housing in the United States and Europe.* Beckenham: Croom Helm.

Holmans, A. E. (1987). *Housing policy in Britain.* London: Croom Helm.

House of Commons Environment Committee. (1982). *First report session 1981-82: The private rented housing sector* (HC401). London: HMSO.

Kemp, P. (1987a). Assured tenancies in rental housing. *Housing Review, 36*(2), 43-45.

Kemp, P. (1987b). The reform of housing benefit. *Social Policy and Administration, 21*, 171-186.

Kemp, P. (1988). New proposals for private renting: Creating a commercially viable climate for investment in rented housing? In P. Kemp (Ed.), *The private provision of rented housing.* Aldershot: Avebury.

Maclennan, D. (1986). *Maintenance and modernisation of urban housing.* Paris: Organisation for Economic Co-operation and Development, Urban Affairs Programme.

National Federation of Housing Associations. (1985). *Inquiry into British housing.* London: Author.

Nevitt, A. A. (1966). *Housing taxation and subsidies.* London: Nelson.

Paley, B. (1976). *Attitudes to letting.* London: HMSO.

Thomas, A. D., with Hedges, A. (1986). *The 1985 physical and social survey of houses in multiple occupation in England and Wales.* London: HMSO.

Todd, J., Bone, M., & Noble, I. (1982). *The privately rented sector in 1978.* London: HMSO.

Todd, J., & Foxon, M. (1987). *Recent private lettings 1982-84.* London: HMSO.

Whitehead, C. M. E. (1983). Housing under the Conservatives: A policy assessment. *Public Money, 3.*

Whitehead, C. M. E., & Kleinman, M. (1986). *Private rented housing in the 1980s and 1990s.* Cambridge: Granta Editions.

Part V

Housing Rehabilitation and Urban Redevelopment

Introduction

JOHN S. ADAMS

GOVERNMENTS OF THE LEFT, right, and center intervene in housing markets on the supply side as well as on the demand side because of the widely held and historically well-documented belief that a private housing market, left to itself, will (a) fail to provide acceptable housing for low- and moderate-income groups; (b) try to maximize profits at the expense of living standards, densities, and construction quality; and, (c) once housing has been built, fail to attract adequate continued reinvestment in long-term maintenance of the stock. A normal outcome is the deterioration of the existing stock. The question of housing rehabilitation and redevelopment of that stock is addressed in the two chapters that follow.

Housing, in addition to being a stock of wealth and a supplier of a flow of housing services, provides society with a large part of the public environment within the settlements where we live. As such, housing is something of a public good from which all people in a settlement derive an aggregate benefit. The government's job is to intervene in the housing market and to adjust the stocks and flows to compensate for private market failures. It is hoped that well-conceived policies and well-designed and -deployed programs will enhance housing environments as a public good, and do so in ways that yield more long-term good than harm to the supply side and the demand side of the housing process.

Housing is a complex system, and counterintuitive outcomes often follow well-intentioned interventions. But interventions are tricky. Rent control, for example, can benefit current tenants but restrict their mobility, penalize owners, raise obstacles to the entry of new residents, curtail new housing production, and promote black-market arrangements in the pricing and allocation of existing stocks. Rent supplements for low- and moderate-income households can lead to higher rents that enrich property owners rather than improve the housing environments for the targeted beneficiaries. Tax breaks for home owners, by stimulating construction and outward migration of central-city residents, can enrich prosperous suburbanites through the above-average property value appreciation they enjoy, but this comes partly at the expense of capital losses endured by the lower-income property owners left behind.

The post-World War II mortgage insurance and home loan guarantee programs, coupled with a separate money market for housing, led to a huge share of U.S. savings going into expanding and improving the American housing stock, but, while this arrangement lowered the cost of housing credit to consumers, it raised the cost of credit to businesses and other borrowers. Thus it seems that an ideal form of public intervention into the housing process has not yet been invented. Each progressive policy and program seems to deliver a flip side of unwanted side effects.

The need for public interventions into housing markets arises regardless of a country's level of development or form of political economy. The need is always present; the differences among countries are in the degrees and forms of responses, and their consequences. In many developing countries, for example, housing has often received a low priority compared with export sectors, the military, public health, education, and other elements of basic physical and human infrastructure. As a result, low- and moderate-income households often are forced to take matters into their own hands, and "self-help" housing is the result, sometimes built by individual families and sometimes by neighborhood and community groups working together.

In many socialist countries, on the other hand, such as those of Eastern Europe and the Soviet Union, the governments consider rental housing an entitlement to be provided at low cost to the consumer. But this policy often leads to waste and shortages, along with allocation methods often unrelated to need or ability to pay, plus a political imperative to speed up production at the expense of quality and durability and to emphasize production of new units at the expense of maintaining the existing stock. It is easy for governments to decide where to allocate housing benefits; what is tougher is deciding how to allocate (or hide, or postpone) the corresponding costs.

One vehicle for housing market intervention on the supply side is the nonprofit community development corporation (CDC), which appeared in Europe and the United States in the middle of the nineteenth century to help meet the housing needs of low-income worker households. At first the housing nonprofits took the form of charitable societies and philanthropic housing trusts. Toward the end of the century, the charitable societies and resettlement agencies in the United States were joined by nonprofit and limited-dividend housing organizations devoted to replacing slum housing.

The modern inner-city CDC appeared in the late 1960s, following the erosion of the antipoverty programs initiated during Lyndon Johnson's administration. CDCs are usually incorporated as nonprofits, although they may have for-profit subsidiaries. They assemble funds from federal, state, and local government loans, from grants from foundations and corporations, and from private gifts and bequests.

No one knows the exact number of CDCs in the United States at the present time, and it is difficult to assess their national impact and their share of housing production and housing improvement. Few if any CDCs have developed to the point that they own or control most of the housing or commercial development in their neighborhoods.

In their study of CDCs in the Cleveland, Ohio, area, Keating, Rasey, and Krumholz found that CDCs have achieved only a limited capacity to address the major problems confronting poor inner-city neighborhoods. Nevertheless, after more than 20 years of existence, the housing CDC movement in the United States, despite its small output to date, has shown itself to be a viable alternative both to for-profit housing produced by speculative developers and to subsidized housing produced by public agencies or by for-profit private developers. The CDC movement must be reformed and reinvigorated, argue Keating and his colleagues, if CDCs are to become and remain major actors in the production of below-market housing in the poor areas of America's inner cities. The activities of CDCs in the production, maintenance, and management of housing in small cities and towns, and in suburban areas—another important subject in its own right—is a topic that lies outside the purview of the study presented in Chapter 12.

In Chapter 13, we learn that three successive conservative governments in the 1970s and 1980s in the United Kingdom have promoted a shift away from direct state intervention in the provision of social services and toward the use of the market and the private sector. Chambers and Gray report how it happened that in the housing arena the government launched a series of bold initiatives based on its commitment to provide for greater private ownership of housing and to substitute private investment for public investment in housing. Production subsidies dropped sharply from previous levels, and more than a million council houses were sold between 1980 and 1987. Meanwhile, consumption subsidies in the form of mortgage tax relief and housing benefits payments have risen.

The pursuit of owner occupation as a cornerstone of housing policy magnifies the question of how to enable low-income owner-occupiers to maintain reasonable housing standards and yet supply the attention and the money to keep the housing stock in good repair, especially as government funding for renewal declines and as the state gradually withdraws from regulating housing standards.

In their review of a succession of programs for renewal and rehabilitation of British housing in the 1970s and 1980s, Chambers and Gray describe how clearance and rebuilding programs were replaced by a series of programs for "target area rehabilitation" with grants for renovation and upgrading of low-quality housing. Their analysis of three Housing Action Area programs revealed no convincing data to support the assumption that the HAA initiatives benefited the most disadvantaged local residents, although they did seem to add momentum to market forces and to facilitate owner occupancy at the expense of private renting. One result was the accentuation of social polarization.

As the recent half century of housing and urban development policy draws to an end, we begin to see old policies and programs discredited and new ones slowly installed to replace them. In the United States and the United Kingdom, the general quality of the housing stocks today is undeniably higher than it was at the end of World War II. The question of whether our housing standards of the last five decades have risen too high or not high enough is continually debated against the press of other national priorities, even while a large share of the households of both countries is better housed than ever before. At the same time, the fractions of our respective populations that are poorly housed in shameful conditions remains unconscionably high. Is this condition properly addressed as a housing problem, or is it better seen as a symptom of the failure of modern society to prepare many of its members with the education, the basic skills, the effective positive role models, the constructive household environments while growing up, the prenatal and postnatal nutrition, the values, and the habits of thought and action that are essential for coping with the complexities of life in our time? Whichever it is, as housing markets are deregulated, programs of housing rehabilitation and urban redevelopment will continue to make a contribution. They cannot do all the job, or even a large fraction of it, but perhaps they can make a significant dent.

12

Community Development Corporations in the United States: Their Role in Housing and Urban Redevelopment

W. DENNIS KEATING
KEITH P. RASEY
NORMAN KRUMHOLZ

IN THE 1980s, THE FEDERAL GOVERNMENT of the United States drastically reduced its role in providing support for subsidized low-income housing. Under the conservative Reagan administration, the budget of the U.S. Department of Housing and Urban Development (HUD) was reduced from $33 billion in 1981 to only $8 billion in 1989. Federal support for local public housing authorities and for local governments in the form of the Community Development Block Grant (CDBG) program was cut back severely. These policies were part of the Reagan administration's attempt to deregulate and privatize the federal government's role in housing as much as possible.

During this period, community development corporations (CDCs) have become important producers of low- and moderate-income housing in the United States. CDCs are neighborhood-based nonprofit organizations chartered to improve their neighborhoods. They can be single purpose (e.g., housing only) or multipurpose (e.g., housing, employment, social services). They are usually tax exempt under the Internal Revenue Service (IRS) Section 501(c)(3) regulations. Many CDCs trace their origins to community advocacy organizations. CDCs are not the only private nonprofit providers of housing; others include churches, labor unions, and special-purpose organizations.

This chapter traces the origins of CDCs, beginning in the late nineteenth century and from the 1960s to the 1980s. It recounts their accomplishments and assesses their impact. It discusses those issues important to the future of CDCs and their role in housing and urban development, and compares them to their European counterparts.

HOUSING AND URBAN DEVELOPMENT IN THE UNITED STATES

Housing and urban development in the United States have been dominated by the private market, including landowners, developers, lending institutions, realtors, and construction firms. Only about 1% of the U.S. housing stock is publicly owned and operated. U.S. public housing occupancy is limited to the poor. Another 1.9 million units are privately owned but publicly subsidized. Finally, over 800,000 private rental units are leased to lower-income renters subsidized by HUD (Leonard, Dolbeare, & Lazere, 1989).

A majority of Americans own their housing, although this proportion declined slightly in 1986, reversing a four decade-long trend toward rising home ownership (National Housing Task Force, 1988). The federal government provides a massive subsidy for home ownership through income tax allowances for mortgage interest and property taxes, which amounted to $54 billion in 1988 (Leonard et al., 1989).

The private home ownership rate in 1986 in the United Kingdom (63%) and Canada (63%) was similar to that in the United States (64%). Canada is similar to the United States, with only 2% of its housing in public rental. In contrast, the United Kingdom had 27% in public rental (local council housing), despite the privatization policies of the conservative Thatcher government. Several other Western European countries have significant proportions of the housing stock in "social housing," either public rental housing or cooperatives, such as Sweden (35%), West Germany (20%), and France (16%) (van Vliet--, 1990).

In the United States, CDCs seek to provide affordable and decent housing to those of low-income with high rent burdens (e.g., those paying more than 50% of their income for rents) and who live in substandard housing. In 1983 an estimated 6 million low-income households paid 50% or more for housing, and 5.5 million lower-income tenants and home owners lived in substandard housing (National Housing Task Force, 1988). These poor households typically are concentrated in urban neighborhoods, often minority in racial composition, with the oldest and most substandard housing.

THE ORIGIN OF NONPROFIT HOUSING

In the United States, the creation of nonprofit housing organizations serving the housing needs of low- and moderate-income households began in the late nineteenth century, when there was great concern about the housing needs of immigrants living in highly dense, unsafe, and unhealthy tenements in urban slums (Lubove, 1962). American housing reformers of this period rejected the policy emerging in Western Europe of municipally owned housing for the working poor. Instead, opposed to municipal "socialism," they favored stricter government regulation of privately owned housing for the poor through housing and health standards embodied in municipal codes (Friedman, 1968). Additionally, some settlement agencies, charitable societies, and private investors and philanthropists joined to form private limited-dividend companies to build new model housing for the poor. Their hope was that private investment would be an alternative to government-subsidized housing and pure philanthropy.

The basic ideas and models for these reformers came from Western Europe, primarily from the British experience (Birch & Gardner, 1981). In response to similar slum conditions, the central governments of several European countries, as early as the 1840s, supported housing nonprofits with loans and later with tax relief to encourage private investment for limited dividends. In 1895 the state of New York, a pioneering state in low-income housing policies, authorized the formation of model tenement societies (Friedman, 1968).

The results of the early American limited-dividend housing societies were mixed. They produced only a few thousand units in a few cities between 1850 and 1920 (Lubove, 1962). Isolated model tenements and their occupants did not transform their surroundings. Many did not house the truly poor for lack of public subsidies. This movement failed to grow because of insufficient private investment due to the limitation on dividends.

FEDERAL INTERVENTION

THE GREAT DEPRESSION

The federal government did not intervene in housing until the Great Depression. Only then and reluctantly did President Franklin Roosevelt's New Deal first create federal home mortgage insurance through the Federal Housing Administration (FHA), and then a federally subsidized but locally administered public housing program (Hays, 1985).

THE GREAT SOCIETY

The modern CDC movement grew out of the antipoverty program conceived by the Kennedy administration and carried out by President Lyndon Johnson's Great Society. Almost from its beginning in 1965, the antipoverty program was plagued by controversy, especially concerning participation by the poor. Advocates of the "empowerment of the poor" sought to incorporate the poor in the decision making of the community action agencies (CAAs) created in inner-city urban neighborhoods. However, this policy was opposed by big-city mayors, who objected to federal funding to local organizations independent of municipal government and to the citizen participation requirements. Their position grew out of serious conflicts between community leaders and city staff in many cities.

This conflict resulted in congressional restrictions on citizen participation and a weakening of federal support for empowerment of the poor. This policy shift was confirmed when the Model Cities program was initiated in 1967 in response to urban ghetto riots. The Johnson administration and Congress agreed that this neighborhood-based redevelopment program was to be controlled by the mayors, with citizen participation to be advisory only.

CDCs can be seen as a response to the failure of the federal government to institutionalize and fully support empowerment of the poor (Perry, 1971). Despite this federal policy, federal support for CDCs was sought and gained through "special impact" funding through the antipoverty program in 1966 (Faux, 1971). The earliest CDCs were funded under this experimental program, terminated in the 1970s.

The urban riots in the summers of 1964-68 resulted in the physical destruction of many urban ghetto areas. The need to rebuild these impoverished areas became a paramount federal concern. In 1968, Johnson's presidential commission on urban housing (the Kaiser Commission) recommended, as part of a massive commitment to a 10-year national housing goal, federal support for nonprofit housing sponsors, including churches and labor unions, to provide subsidized housing for the poor. Since 1937, federally subsidized housing for the poor had been provided by local public housing authorities. Beginning in 1961, private, for-profit developers received development subsidies (e.g., below-market mortgages) with limited dividends in return for providing occupancy for low- and moderate-income residents. The only federal subsidy program for nonprofits was a small elderly housing program created in 1959.

The Kaiser Commission (1968) rejected public housing as a viable program for providing the greatly expanded production of below-market housing that it recommended. Instead, it supported the use of federal tax incentives and subsidies necessary to obtain the involvement of for-profit developers, lenders, and investors. In landmark 1968 housing legislation, Congress established this type of housing subsidy program, which also allowed for the participation of nonprofit sponsors. Congress created the National Corporation for Housing Partnerships as a vehicle for corporate investment in below-market housing at a competitive profit rate.

CONSERVATIVE REACTION: AUSTERITY AND RETRENCHMENT

The 1968 election of conservative President Richard Nixon did not immediately lead to the end of these liberal Democratic initiatives. In its first term, the Nixon administration mostly carried out the national housing goals and continued the antipoverty and Model Cities programs. However, in Nixon's short-lived second term his administration attempted to dismantle these programs and imposed a moratorium on federal housing subsidy programs in 1973 on the grounds that they were inefficient and inequitable failures. In 1974 a Community Development Block Grant program was created to replace several federal housing and development programs. A new rent subsidy program—Section 8—was also created to provide subsidies to low-income beneficiaries, enabling nonprofit and for-profit developers to build and rehabilitate below-market housing for occupancy by subsidized tenants. CDCs have used Section 8 subsidies extensively in their projects, and many have received operating and project subsidies through CDBG funding from local governments.

EXPANSION OF THE CDC MOVEMENT

The decade 1965-75 was a period in which CDCs were created, received federal, foundation, and corporate assistance, and gained their initial development experience. During the 1970s the CDC movement grew as an offshoot of community protests.

Many inner-city neighborhoods protested against mortgage and insurance "redlining" by the FHA and private lenders and insurers. They persuaded Congress to enact two important federal antiredlining laws governing federally regulated lenders. The Home Mortgage Disclosure Act (HMDA) of 1975 and the Community Reinvestment Act (CRA) of 1977 pressured lending institutions to invest in housing in older urban neighborhoods where CDCs formed. This legislation was extended and strengthened by Congress in 1988.

During the 1970s, Neighborhood Housing Services (NHS) was created. NHS represents partnerships of local lenders and community groups to provide mostly "conventional" private mortgage financing for housing in formerly redlined neighborhoods. This has made CDC housing programs more viable. NHS is now supported by the Neighborhood Reinvestment Corporation, representing several federal agencies involved in housing and regulation of lenders.

The Carter administration undertook several initiatives aimed at promoting neighborhood development. It created a Neighborhood Self-Help Development (NSHD) program to provide direct funding for CDCs. It also created the Urban Development Action Grant (UDAG) program, which has been used in many cities for neighborhood development projects, including housing. Finally, in 1979 Carter appointed a National Commission on Neighborhoods, which recommended greatly increased support for neighborhoods and neighborhood-based development. While the Democratic Carter administration was much more supportive of CDCs than the Republican administrations of Nixon, Ford, and Reagan, it did not support the recommendations of its own National Commission on Neighborhoods and provided only modest funding for CDCs.

With Carter's defeat and the advent of Reagan's much more conservative "new federalism" in 1981, even this modest federal support for CDCs and below-market housing subsidy programs was drastically reduced. The NSHD and UDAG programs were eliminated, and the CDBG, Section 8, and public housing programs were cut drastically. In the face of these federal budgetary cutbacks, CDCs were forced to seek other sources of financial support.

Private philanthropic foundations, both national and local, have provided support for CDCs. In 1980, the Ford Foundation created the Local Initiatives Support Corporation (LISC) to solicit corporate support for CDCs. During 1980-87 LISC's grants and loans to CDCs totaled $72 million, resulting in the development of 12,950 housing units (LISC, 1987). In 1988, LISC also raised $51 million for a national equity fund partnership for CDCs. In 1981, developer James Rouse created the Enterprise Foundation to promote neighborhood-based housing for the urban poor. The Enterprise Social Investment Corporation (ESIC) provides equity capital to CDCs through syndication of the low-income housing tax credits created in the 1986 federal tax reform act. By 1987, ESIC had assisted the development of 2,000 units (Enterprise Foundation, 1987). In 1988, the National Council on Community-Based Development was formed to increase private sector support for CDCs.

While increased support provided by state and local governments and philanthropic and corporate funders has been important to the survival and expansion of CDCs, this has not made up for the substantial loss of federal support for low-income housing since 1981. By 1989, HUD had virtually no new funding for new and rehabilitated low- and moderate-income housing in the United States. The federal low-income housing tax credits have attracted only modest corporate investment and have produced low-income housing in tandem with other direct subsidies.

OPERATION AND IMPACT OF CDCs

The operation of CDCs varies widely in the United States. The scale and impact of their operation depend upon their purposes, histories, service areas, funding, and leadership. A 1988 national survey identified 694 CDCs that have completed housing projects. The 631 that provided data reported that they had produced 124,938 housing units. Approximately 35% represented newly constructed units and the remainder was rehabilitation. In addition, 290 of these CDCs reported doing home repairs and weatherization on another 275,062 units (National Congress for Community Economic Development, 1989).

The evaluation standards for judging the performance of CDCs are not easy to define or quantify (Garn, 1975). The typical measures of success are the number of housing units built and rehabilitated and the amount of investment, both public and private, generated by the CDC. A less tangible measure is the contribution of the CDC to neighborhood stability and revitalization.

An early assessment of CDCs funded under the antipoverty program found that most CDCs were successful (Abt Associates, 1973). An evaluation of 100 CDCs that participated in HUD's NSHD program concluded that generally they successfully implemented their planned projects. Key factors in their success included skilled staff, strong leadership, and sufficient external support (Mayer, 1984). CDCs have made their greatest impact in cities like Boston, Chicago, Cleveland, New York, and San Francisco, where citywide CDC networks have emerged to enhance fund-raising, technical assistance, and training (Bratt, 1989).

GOALS OF CDCs

In their infancy, some saw the primary goal of CDCs as virtually complete political and economic independence and self-sufficiency (Hampden-Turner, 1975). If this is the criterion by which we judge CDCs, then they have failed. No CDC has developed to the extent that it entirely owns or controls the housing and land within its territorial boundaries and is self-sufficient.

CDCs have not attained political autonomy, as envisioned by those who advocated neighborhood self-government (Kotler, 1969). However, most CDCs do not have such ambitious goals (Sturdivant, 1971). To the contrary, most CDCs are largely apolitical, even though many trace their origins to grass-roots community groups that engaged in political protest and electoral politics. CDCs are typically prohibited by their own charters and IRS regulations from engaging in politics and lobbying.

Contrary to the ideals of empowerment of the poor, many CDCs have narrow membership bases, and most are directed more by their staffs and boards of directors than directly by the residents of their communities. Pecorella (1984) has termed this the process of "professionalization" of CDCs. He concludes that CDC leaders still encourage citizen participation, at least in setting broad goals and policies for neighborhood revitalization, but not nearly as much in the implementation of projects and programs. This view contradicts the goal of widespread grass-roots citizen participation espoused originally by advocates of empowerment of the poor (Kanter, 1971).

Since CDCs are mostly engaged in development projects, they have become less likely to be engaged in community organizing (Fulton, 1987). It has been argued that CDCs cannot simultaneously act as developers, property owners, and managers and also organize community residents, because this might involve them in conflict with their own constituents over such potentially controversial issues as the selection of owners and tenants in CDC-owned and -managed housing projects, the setting of rents, and evictions (Clarkson, 1987). It has also been suggested that CDCs may be co-opted from goals such as social, nonmarket housing if they become reliant upon funding from corporate and government sources or begin to act like private, profit-motivated real estate developers. However, Bratt (1986, 1987) argues that CDCs have avoided taking on the attributes of the private market, that they have remained socially conscious in their goals, and that they remain at least small-scale agents of progressive, equitable social change in the housing market as nonprofit developers.

In view of the greatly reduced federal support for public housing and the emphasis upon the role of the private sector in providing lower-income housing, CDCs have an important role to play. The for-profit sector has shown relatively little interest in the development of low-income housing except for passive participation in tax-sheltered investment. CDCs present the best "private" alternative to public authorities in the provision of low-income housing in the United States. While it is unclear just how great the capacity of CDCs is for making larger contributions to the supply of affordable housing, it is clear that mature CDCs with proven records in housing could build and rehabilitate much more housing if they were assured of increased financial support, both for their operations and for their housing projects (Bratt, 1986; Mayer, 1988). The support that CDCs have received from private foundations (e.g., LISC and Enterprise) and the corporate sector since 1980 has been important in sustaining and expanding their operations. However, the private sector support that CDCs received in the 1980s was minuscule compared to the loss of federal housing subsidies under the Reagan administration.

The only CDC initiative of the Reagan era was forced upon HUD by Congress (as were the few other small new federal housing programs during this era)—the Neighborhood Development Demonstration Program. Only a token $1 million was appropriated, and only 38 CDCs in the United States received funding, although even the use of this assistance proved successful (Pratt Institute Center for Community and Environmental Development, 1988).

It is clear that CDCs must receive greatly increased support from the federal government if they are to expand. The National Housing Task Force (1988), commissioned by Congress to recommend new approaches, points to the need for federal support for national, regional, and local nonprofits and their housing programs as an important component of a new national housing policy. Beset by scandals surrounding private exploitation of federal housing subsidy programs, HUD in 1989 had not addressed the role of CDCs in federal housing policy. HUD Secretary Jack Kemp is a well-known champion of privatization. A liberal Democratic proposal titled the Community Housing Partnership Act has proposed $10 million in "capacity-building" funding for CDCs and $500 million for CDC development of low-income housing annually, distributed by HUD to state and local governments for allocation to CDCs (Dreier, 1987). If adopted, this would establish a federal policy recognizing the role of CDCs but would not provide nearly enough funding to enhance their housing production capacity significantly. CDCs will remain dependent on a combination of public and private funding generated at national, state, and local levels.

CONCLUSION

After two decades of existence, the housing CDC movement in the United States has established itself as a potential alternative both to for-profit housing produced by speculative developers and to state and local public housing authorities. To date, it has produced only a relatively small amount of low-income housing, but it has the potential to provide much more if adequate support is provided. (See the Appendix to this chapter for a list of national support organizations for CDCs in the United States.)

While CDCs have not attained the goals of empowerment of the poor or neighborhood self-reliance advocated by their more radical and utopian advocates in the 1960s, modern CDCs have outlived and out-performed their nineteenth-century counterparts. However, like those earlier limited-dividend housing companies, inner-city CDCs in the late twentieth century have been unable on their own to transform troubled poor neighborhoods. CDCs have contributed to the stability and at least limited revitalization of many neighborhoods. In those neighborhoods where "gentrification" has occurred, CDCs are often the only hope for the preservation of affordable housing for longtime and lower-income residents.

Some have feared that CDCs reliant upon private support and acting like developers would become subsumed by the private market and lose their social goals. There is no evidence to suggest that CDCs have compromised their goals of serving low-income residents. On the other hand, as Bratt (1986) admits, CDCs in the United States have admittedly reformist goals, must work within the existing private market system, and have a limited capacity to produce nonmarket housing in those neighborhoods in which they operate. Since their primary focus has been on individual housing projects, few CDCs have become involved in comprehensive neighborhood planning. CDCs have proven to be largely apolitical and depend for their support upon the largess of public and private funders whom they cannot afford to alienate.

Yet, without a strong political base, it is unlikely that CDCs will be able to win the support that they need nationally and locally. Otherwise, their fate may be to become only marginal producers of small amounts of below-market housing in the poor neighborhoods of a scattering of cities. Eventually, it may become difficult to sustain many CDCs under these circumstances. CDCs or affiliated neighborhood-based organizations must reorganize a base of political support that legitimates their role and provides long-term support.

In building a more potent political coalition, CDCs must develop their own progressive housing programs. CDC leaders have recognized the limitations and drawbacks of reliance upon investment by corporate supporters attracted by the federal low-income tax credit, which is a complex and cumbersome method of attracting equity capital. If corporate and philanthropic goals should shift or matching public funds prove inadequate, then CDCs may lack the necessary operating resources. While they may derive income from the management of their properties, this is unlikely to prove adequate to sustain their operations.

Therefore, the government must be persuaded to support nonprofit housing and CDCs to a much greater extent than it has in the past. A visionary version of national legislation has been developed (see Institute for Policy Studies Working Group on Housing, 1988), but it lacks a visible and powerful national constituency. Developing this political support poses a major challenge for CDCs in the 1990s. It is unlikely that there will soon be in the United States a nonmarket "social housing" sector comparable to what exists in many Western European countries. What may evolve is a viable and expanding CDC housing movement that is an alternative to the private market that dominates housing in the United States.

APPENDIX:
NATIONAL SUPPORT ORGANIZATIONS
FOR CDCs IN THE UNITED STATES

- Center for Community Change
  1000 Wisconsin Avenue, NW
  Washington, DC 20007
- Council for Community-Based Development
  127 East 28th Street
  New York, NY 10016
- The Enterprise Foundation
  505 American City Building
  Columbia, MD 21044
- Local Initiatives Support Corporation
  666 Third Avenue
  New York, NY 10017

- National Congress for Community Economic Development
  1612 K Street, NW, Suite 510
  Washington, DC 20006
- National Reinvestment Corporation
  1325 G Street, NW, Suite 800
  Washington, DC 20005

REFERENCES

Abt Associates. (1973). *An evaluation of the Special Impact Program.* Cambridge, MA: Author.

Birch, E. L., & Gardner, D. S. (1981). The seven-percent solution: A review of philanthropic housing 1870-1910. *Journal of Urban History, 7,* 403-438.

Bratt, R. G. (1986). Community-based housing programs: Overview, assessment, and agenda for the future. *Journal of Planning Education and Research, 5,* 164-177.

Bratt, R. G. (1987). Dilemmas of community based housing. *Policy Studies Journal, 16,* 325-334.

Bratt, R. G. (1989). *Rebuilding a low-income housing policy.* Philadelphia: Temple University Press.

Clarkson, J. (1987). From organizing to housing development: It's tough wearing two hats. *Shelterforce, 10*(3), 14-17.

Dreier, P. (1987). Community-based housing: A progressive approach to a new federal housing policy. *Social Policy, 18*(2), 22.

Enterprise Foundation. (1987). *Annual report.* Columbia, MD: Author.

Faux, G. (1971). Politics and bureaucracy in community-controlled economic development. *Law and Contemporary Problems, 36,* 277-296.

Friedman, L. M. (1968). *Government and slum housing: A century of frustration.* Chicago: Rand McNally.

Fulton, W. (1987). Off the barricades, into the boardrooms. *Planning, 53*(8), 11-15.

Garn, H. A. (1975). Program evaluation and policy analysis of community development corporations. In G. Gappert & H. M. Rose (Eds.), *The social economy of cities* (pp. 561-588). Beverly Hills, CA: Sage.

Hampden-Turner, C. (1975). *From poverty to dignity.* Garden City, NY: Anchor.

Hays, A. R. (1985). *The federal government and urban housing: Ideology and change in public policy.* Albany: SUNY Press.

Institute for Policy Studies Working Group on Housing. (1988). *The right to housing: A blueprint for housing the nation.* Washington, DC: Author.

Kaiser Commission. (1968). *A decent home.* Washington, DC: Government Printing Office.

Kanter, R. M. (1971). Some social issues in the community development corporation proposal. In C. G. Benello & D. Roussopoulos (Eds.), *The case for participatory democracy: Some prospects for a radical society* (pp. 64-71). New York: Viking.

Kotler, M. (1969). *Neighborhood government: The local foundations of political life.* Indianapolis: Bobbs-Merrill.

Leonard, P. A., Dolbeare, C. N., & Lazere, E. B. (1989). *A place to call home: The crisis in housing for the poor.* Washington, DC: Center on Budget and Policy Priorities and Low Income Housing Information Service.

Local Initiatives Support Corporation. (1987). *Annual report.* New York: Author.

Lubove, R. (1962). *The progressives and the slums: Tenement house reform in New York City, 1890-1917.* Westport, CT: Greenwood.

Mayer, N. S. (1984). *Neighborhood organizations and community development: Making revitalization work.* Washington, DC: Urban Institute Press.

Mayer, N. S. (1988). *The role of non-profits in renewed federal housing efforts.* Cambridge: MIT Center for Real Estate Development, MIT Housing Policy Project.

National Congress for Community Economic Development. (1989, March). *Against all odds: The achievement of community-based development organizations.* Washington, DC: Author.

National Housing Task Force. (1988). *A decent place to live.* Washington, DC: Author.

Pecorella, R. T. (1984). Resident participation as agenda setting: A study of neighborhood-based development corporations. *Journal of Urban Affairs, 6,* 13-27.

Perry, S. E. (1971). A note on the genesis of the community development corporation. In C. G. Benello & D. Roussopoulos (Eds.), *The case for participatory democracy: Some prospects for a radical society* (pp. 55-63). New York: Viking.

Pratt Institute Center for Community and Environmental Development. (1988). *An evaluation of the Neighborhood Development Demonstration.* Washington, DC: U.S. Department of Housing and Urban Development.

Sturdivant, F. D. (1971). Community development corporations: The problem of mixed objectives. *Law and Contemporary Problems, 36,* 35-50.

van Vliet--, W. (Ed.). (1990). *The international handbook of housing policies and practices.* Westport, CT: Greenwood/Praeger.

13

Housing Rehabilitation in Contemporary Britain: The Evolving Role of the State

DAVID CHAMBERS
FRED GRAY

OVER THE LAST DECADE, a number of areas of social provision in Britain have moved along the public-private continuum. This shift has been away from direct state intervention, particularly in allocation and funding, and toward market criteria and the private sector. It has been guided and overseen by the three Conservative governments led by Prime Minister Margaret Thatcher. One of the ironies of the situation is that the movement to assert the importance and value of the private sector in social provision has required a considerable legislative program and set of policy initiatives. There has also been considerable political argument and debate about this aspect of Thatcherism, particularly in the fields of health care and education (Ball, Gray, & McDowell, 1989).

It is in housing that the most radical and significant change of direction has taken and is taking place. Most obvious is the manner in which the state social housing sector has come under attack (Forrest & Murie, 1988). In addition, various initiatives have emerged aimed at reinforcing the government's commitment to the private ownership of housing and substituting private for public investment (Crook, 1986). This is exemplified by direct government expenditure on housing, which decreased by 60% (in real terms) between 1978-79 and 1987-88. The withdrawal of production subsidies is best evidenced by the slump in new public sector building. When combined with a vigorous council house sales program (more than a million properties sold between 1980 and 1987; Central Statistical Office, 1988), this helped reduce the proportion of dwellings in public ownership from 31.5% in 1979 to 25.9% in 1987 (Central Statistical Office, 1989). A note of caution is warranted in using official housing expenditure statistics. A broader definition of housing expenditure including mortgage tax relief and housing benefits payments—both consumption subsidies, often to the private sector—reveals an increase rather than a decrease in state expenditure over the same period.

This chapter is concerned with a less apparent dimension of housing reform: the manner in which public intervention in private sector housing rehabilitation has evolved over time and is at present being restructured. This restructuring is not peripheral to but a significant part of the drive toward expanding owner occupation. As Crook (1986) observes, "Government attention has been directed at ways of subsidizing the repair and upgrading of [these] houses through the provision of grants and publicly sponsored area-revitalisation programs."

The pursuit of owner occupation, as a cornerstone of popular capitalism, is not without its problems. In particular, issues have emerged surrounding how to enable low-income owner-occupiers to maintain reasonable housing standards and to keep the stock in good repair (Brindley & Stoker, 1988, p. 48). In this context housing rehabilitation stands at a watershed.

Our focus is on housing rehabilitation and the manner in which central and local governments have attempted to secure certain changes in the private housing market. In part, recent changes in urban renewal in Britain have occurred by stealth: Funding for renewal has been cut back, at least from the mid-1980s, and there has also been a more gradual withdrawal from state regulation of housing standards. Running parallel with such changes has been an increasing emphasis on solutions to private sector renewal problems found within the private sector itself. Policy analysis has ranged across a variety of issues, including how to fund maintenance (Brindley & Stoker, 1988) and how best to involve householders (Kintrea, 1987; Thomas, 1985).

One of the problems in perceiving the extent and character of recent shifts in this dimension of housing policy stems from the fact that urban rehabilitation in Britain has shifted in its objectives and characteristics throughout this century. The evaluation of policy shifts is therefore difficult because of the absence of an obvious starting point. Doling and Stafford's (1989) assertion that there has been a divorce between policies devoted to housing access on one hand and maintenance and rehabilitation on the other also helps account for the weak understanding of many important policy relationships. Indeed, only now, after a decade of Thatcherism, is the extent of contemporary change becoming apparent. Our concern is with two specific dimensions of housing rehabilitation policy: first, proposals aimed at reshaping inner-city housing rehabilitation in Britain, and second, aspects of the Conservative government's housing policy that seek to establish "partnerships" between local government and the private sector in achieving this objective. Given the incremental nature of British housing policy, any discussion or assessment of current proposals for change needs to be set within the context of what has gone before. It is therefore important to provide a historical perspective to these emerging policy issues. To this end, the approach adopted here first takes a retrospective view of interventionist approaches to private sector housing and progresses to an analysis of the prospects for the sector as we enter the last decade of the century. For both sections we provide more detailed examples of policies in action in specific locales.

BACKGROUND

The "why" of public intervention in the private housing market has been debated at length in the context of both its production and its consumption. This debate is not addressed here, although, as Thomas (1986) observes, while the form of public sector involvement has varied over time, its necessity stems from the fact that

> if left to itself, the private housing market will fail to provide acceptable conditions for low-income groups; will attempt to maximise profits at the expense of space standards, densities and building quality; and, once housing has been built, will fail to support adequate investment in maintaining the stock. (p. 1)

Three basic principles underpin this raison d'être of state intervention, accepted by successive governments whatever their political complexion, and have become inculcated in British housing legislation: the concept of fitness of housing for human habitation, the availability of government grants for the improvement and rehabilitation of existing private sector dwellings, and the concept of intervention, on the basis of discrete geographical areas and over short time scales, to secure the improvement of the housing stock. Although our primary concern is with area improvement policies, it is important to stress the centrality of the concept of "fitness for human habitation." The concept defines the minimum acceptable standard of accommodation below which the state will intervene to secure improvement. This interventionist standard underpins the clearance and rehabilitation programs discussed below.

INTERVENTION STRATEGIES

Against this background, three distinct but overlapping phases of state intervention may be identified. First, an initial phase of clearance followed by municipal redevelopment dominated state policies in the period 1930 to the mid-1960s (English, Madigan, & Norman, 1978; Gibson & Langstaff, 1982; Thomas, 1986). The criterion applied in selection for clearance was whether a significant proportion of dwellings in a given area fell below the minimum standard of fitness coupled with an acknowledgment that it was impractical, in economic terms, to secure improvement of existing dwellings.

The break with this approach in the 1960s has been attributed to both a dissatisfaction with the destruction of existing communities (Cockburn, 1977; Doling & Stafford, 1989; Young & Willmott, 1957) and the imperatives of one of Britain's recurrent economic crises. The legislative potential for a move away from clearance was made possible in 1964, following the introduction of Improvement Areas in the Housing Act of that year. Improvement Areas were to be designated in small neighborhoods where more than half of the houses lacked at least one of the basic amenities, which included a lavatory and bathing facilities. The improvement of properties within these areas was to be grant aided.

The approach proved markedly unsuccessful, in part because of the lack of government support and the continued commitment by local government to comprehensive redevelopment. Indeed, Spencer's (1970) analysis of early housing improvement policies concludes that within the 400 Improvement Areas declared between 1964 and 1968, only 4,000 houses were improved.

The second phase of state intervention is more clearly marked by the introduction of the General Improvement Area (GIA) in the Housing Act of 1969. The major objective of this approach was to focus on the rehabilitation of dwellings specifically in areas comprising poor-quality housing but that had, subject to timely intervention and improvement, the potential of providing homes for an extended period. Greenwood (1969) succinctly states the mood of the time:

> It [had] become increasingly clear that it would make economic sense to prevent old houses deteriorating to slums and to avoid, as far as possible, the breaking up of a well established community by the clearance and redevelopment of a whole area.

The period 1969 to 1974, dominated by the GIA approach, is much discussed in the literature (Roberts, 1976; Thomas, 1986). This phase of intervention was again linked to the principle of a benchmark of fitness for habitation. The success of the policy was dependent on the availability of house renovation grants, funded by central government but administered by local government. This phase of intervention is important because it was marked by an increasing acceptance by government that it was right and proper for it to become involved in the repair of properties, a role previously left to property owners themselves (Doling & Thomas, 1982). In much the same way as clearance waned as a consequence of dissatisfaction with its impact, so there was mounting criticism of the failure of GIAs to reach the right people or the right houses (Cullingworth, 1979).

The tendency for GIAs to avoid the worst areas of housing was emphasized by publication of the National House Condition Survey 1971, which demonstrated the scale of the substandard property problem. The survey identified some 1.2 million unfit dwellings in England and Wales (7.3% of the total) and approximately 2.8 million dwellings lacking all of some of the basic amenities.

The government's response, published in two white papers (H.M. Government 1973a, 1973b) recommended far-reaching changes in the machinery for public intervention into the private sector market. This heralded the start of the third phase of intervention. The initiative to emerge was the Housing Action Area (HAA), a second-generation area rehabilitation approach. While retaining some of the existing aims (physical improvement of the housing stock) HAAs stressed social objectives. Essentially, the new approach, although retaining the framework of intensive activity in a small geographical area, acknowledged the potentially socially disruptive nature of proactive rehabilitation. It had at its heart the interests of the people who lived in the area, and the desire to maintain existing communities (Channon, 1974).

Despite the fall of the Conservative government in 1974, the concept of HAAs was carried forward by the newly elected Labour administration and, with minor amendments, was finally introduced in the Housing Act 1974. For a decade after their introduction, HAAs were a dominant form of state intervention in private sector rehabilitation. It is notable, however, that despite their durability as a policy, assessment of their impact is remarkably limited. Such research as has been reported is characterized by analysis of data at the national level and over limited time scales (H.M. Government, 1976, 1978) or highly specific data relating to individual HAAs, traditionally published in journals with a professional bias (see Cameron & Smart, 1982; Henney, 1976; Rowland, 1984; Turney, 1980). Both types of research limit the extent to which generalizations can be made as to the success or failure of the approach.

Until the mid-1980s, this third phase of intervention was marked by bipartisan political acceptance. This is evident in both its passage into legislation and subsequent implementation by local authorities of differing political complexions and support by Labour and Conservative governments (Chambers, 1988). For example, although HAAs received most attention by local authorities in the five-year period immediately after their introduction, the 1979-83 Conservative government continued the funding of house renovation grants and, indeed, at enhanced rates within HAAs. In the context of Thatcherism, however, this considerable public intervention in the private market was something of a paradox. Given the Conservative government's dislike of public, and more especially local authority, intervention and public housing, a realistic conclusion is that private sector house renovation, and hence area improvement, was the acceptable face of public expenditure, acting to support and subsidize the private market.

Such support for the private sector was clearly evidenced by the relaxation of local authority expenditure on house renovation grants during the financial years 1982-83 and 1983-84. Indeed, despite massive cutbacks in most areas of local authority housing expenditure, spending on renovation grants increased in England and Wales from £101 million in 1978-79 to £1,013 million in 1983-84 (H.M. Government, 1985). It is perhaps significant that this boom period for house renovation grants coincided with the government's explicit drive to encourage owner occupation and an impending general election. It is less than coincidental that the scale and success of this expenditure eventually prompted treasury concern. The result, by the mid-1980s, was a policy U-turn—stemming the flow of the house renovation grant tap and radical rethinking about the future of HAAs and home improvement policies.

These major shifts in the Thatcher government's renovation policies are well illustrated by the changing number of home improvement grants in England, which rose from 70,600 in 1979-80 to 250,600 in 1983-84 and then fell to an estimated 111,000 in 1988-89 (H.M. Government, 1989a).

In May 1985, the Secretary of State for the Environment presented to Parliament a Green Paper proposing changes in home improvement legislation. The philosophy underlying the green paper's proposals, while retaining the major tools of urban rehabilitation (area improvement of discrete geographical areas and house renovation grants), was marked by a fundamental shift in emphasis. The core of the proposals was that state intervention into private sector housing rehabilitation was justified, but the primary responsibility for maintaining and improving property should rest with its owner (or occupier) in both the owner-occupied and private rented sectors alike. It was recognized, however, that in certain circumstances building societies, other financial institutions, and local authorities should assist owners in meeting those responsibilities. This stance appears to have stemmed partly from a belief that grant monies may have been going to people who could adequately afford to undertake improvements at their own expense. However, no evidence to substantiate this view emerged from the government's Distribution of Grants Enquiry, conducted from 1981 to 1983.

The Green Paper proposals, by the late 1980s, were being translated into fresh legislation, founded on fundamental changes in the existing policy frameworks for area improvement and the basic fitness standard, on which this framework rests. Taking the latter aspect first, the proposed amendments to the standard of fitness imply a reduction in the future level of intervention. Without any increase in the number of properties renovated, the definitional changes alone would reduce the number of dwellings classified as unfit. Ormandy (1985) sees this proposed revision as the central plank of the government's strategy.

On the question of house renovation grants, Gray and Chambers (1985), in a critique of the Green Paper, conclude that the proposals would progressively dismantle the present grant system. This change stems from the government's dilemma of a need for state intervention to improve bad housing while simultaneously accepting that public expenditure should be minimized.

Of central importance here is the continued belief in an area improvement approach (H.M. Government, 1985). Clearly, any new area approach would, however, need to operate within the constraints imposed by a more restrictive renovation grant policy. To this end the government intends to make the future financing of area rehabilitation less reliant on exchequer monies. A new statutory area initiative is to be introduced, with local authorities acting as coordinators and enablers by securing private funding. This is clearly spelled out in the government's recent expenditure plans, which confirm that the Government aims to stimulate *private spending* so as to ensure that the housing stock is kept in good repair, and to concentrate public sector resources on those people unable to meet the full cost of repairs and on those areas where problems are most severe. Proposals for legislation in the 1989-90 Parliamentary Session will secure the targetting of home improvement grants on those people who will be unable to finance the necessary work themselves . . . help elderly people with minor repairs; encourage authorities to take a more balanced approach to clearance and renovation, and create a single type of statutory renewal area. (H.M. Government, 1989a, para. 35; emphasis added)

It is significant that these radical reforms—the reformulation of the fitness criterion, reduction of public funding, and remoulding of area intervention—are not based on any empirical investigation or verification of the inadequacies or otherwise of existing or proposed rehabilitation policies. Given the government's assumption that area action remains an appropriate mechanism to secure housing rehabilitation and the absence of evidence to support this view, it is important to explore the consequences of existing area improvement policies. In particular, to what extent and how does the resident population benefit from area rehabilitation, and what are the effects of such policies on tenure patterns? The following discussion considers the results of recent research (Chambers, 1988), on the impact of three large HAA programs conducted by the London boroughs of Hammersmith and Fulham, Islington, and Lambeth. This research also explores the attempt in Hammersmith and Fulham, by the local authority, to secure private funding for area improvement.

AREA IMPROVEMENT POLICIES IN ACTION: THE EMPIRICAL EVIDENCE

The thrust of area improvement policies since their introduction in 1964 has centered on the physical improvement of housing as an alternative to clearance and redevelopment. It was only in the second-generation approach, the HAA, that there was an explicit recognition of the need to improve the housing stock of an area for the benefit of the existing inhabitants. Accordingly, one measure of success in HAA terms should be the extent to which existing communities have been retained but in better housing conditions. However, minimal data have been gathered on whether or not this has occurred. Certainly housing has been improved, but the extent to which local people have benefited is unclear. In the absence of such crucial information, it is difficult to accept uncritically that the previous history of area intervention warrants its retention as the template for a third-generation approach.

Information on HAA progress available to date has been flawed either by its aggregation and analysis at a national level (which limit its usefulness in commenting on local influences in policy outcomes) or by the adoption of a case-study approach that considers individual HAAs (this tends toward anecdote, with limited possibilities for generalization). More important, both strategies have a tendency to rely on data taken at a single point in time, an approach that revolves around relatively easily monitored indicators, such as number of dwellings improved, and fails to incorporate dynamic features of an interventionist policy, such as population turnover.

Recently, Chambers (1988) has made available a time-series analysis of the HAA programs of three London inner-city boroughs, chosen for their similarities in terms of housing conditions (including age of stock, state of repair, and level of overcrowding) and the size of the programs pursued by them since 1974. In total, 45 HAAs were declared within the authorities, containing between them some 21,000 dwelling units. The following discussion is based on these data, which were obtained from council reports and other local authority documentation.

The evidence to emerge indicates that intervention in HAAs has been characterized by dramatic changes in tenure redistribution. The two important common trends to emerge for all three boroughs are a major decline in private renting and concomitant increases in owner occupation.

It could be argued that this picture of change merely mirrors both boroughwide changes within all three authorities and the general pattern of tenure redistribution for Britain (Murie & Malpass, 1987) during the 1970s and 1980s. However, the pace of change within the studied HAA programs has been accelerated in consequence of this intensive form of intervention. Our argument, then, is that HAA policies facilitate and heighten general market trends.[1]

Table 13.1 aggregates data for all the individual HAAs within the three studied boroughs. Disaggregating the data reveals more complex patterns within and between individual HAAs and boroughs. In Hammersmith and Fulham, reductions in private rental stock within individual HAAs ranged from 48% to 6% of the total stock over the five-year, or seven-year, term of activity.[2] In Islington, detailed moni-

TABLE 13.1

Tenure Redistribution in Housing Action Area Programs in Three London Boroughs

Tenure	Start	Finish	% Change[a]
Hammersmith and Fulham			
owner occupation	2,882	4,824	+13.9
private rented	8,400	4,917	−25.4
local authority	1,140	1,575	+ 3.1
housing associations	1,376	2,547	+ 8.4
all tenures (dwellings)	13,798	13,863	
Islington			
owner occupation	1,104	1,514	+10.3
private rented			
resident landlord	866	511	− 8.8
absentee landlord	1,260	501	−19.0
local authority	495	1,111	+15.4
housing associations	274	355	+ 2.0
all tenures (properties)	3,999	3,992	
Lambeth			
owner occupation	764	1,368	+10.1
private rented	1,408	1,039	−16.2
local authority	945	1,204	+ 1.1
housing associations	219	615	+ 8.2
other tenures (service tenancies)	127	24	− 3.0
all tenures (dwellings)	3,463	4,250	

NOTE: The data for the three boroughs are the aggregated information for individual HAAs derived from council reports and other documentation. Differences in tenure and property/dwelling classification reflect individual monitoring procedures.
a. Change as a percentage of the total HAA housing stock.

toring distinguished between resident and absentee landlords. Absentee landlords declined in the range of 33% to 8%, while resident landlords fell in the range of 13% to 6%. In Lambeth there was a reduction, in individual HAAs, in the private rented sector in the range of 3% to 28%.

While a reduction in the private rented sector is common to all three authorities, the pattern of expanding tenures is less consistent. In Hammersmith and Fulham HAAs, owner occupation as a proportion of the total stock increased in the range of 33% to 0. The variability of tenure shifts in the public and housing association tenures ranged from 22% to −4% and 35% to −4%, respectively. Islington, by comparison, showed increases in public sector stock from 40% to 1%. In Lambeth, the public sector changed less dramatically, in the range of −4% to

18%, although there was a more marked shift into the housing association sector. These differences are attributed to changes in central/local relations, local political priorities, and the ebb and flow of government funding for renovation grants and compulsory purchase.

The analysis of HAAs by individual local authorities notably lacked any meaningful attempt to monitor population movements into and out of HAAs. For this reason, it was impossible for authorities to say whether or not HAAs benefited people resident at the time of an HAA's declaration. However, in Islington it was possible to establish from limited and incomplete data that 1,243 persons, or 64% of those re-housed during the life of an HAA, moved out of the area of the HAA in question. Islington Council made public no information on whether moves from the HAAs were voluntary or forced. In addition, the available evidence suggests that the HAA policy of all three authorities left private rented dwellings with the worst housing problems. For example, one consistent statistic that emerges is that, at the completion of an HAA, poor conditions had become residualized within this tenure (Chambers, 1988). The validity of area-based approaches with the objective of improving housing for the benefit of existing inhabitants therefore appears questionable.

PROSPECTS

The future of public intervention in private sector housing rehabilitation is constrained by two contradictory imperatives. On the one hand, the inadequacies of private housing market processes emphasize the need for continued government activity. On the other, a commitment to contain and curtail public expenditure militates against the former. In effect, the stance taken by government in 1985 remains its current priority, although now more explicitly stated. In September 1987 the government published its housing policy intentions, which ranged across both the public and private sectors (H.M. Government, 1987). These were incorporated, in part, into the Housing Act of 1988, with further legislation in progress in 1989 (H.M. Government, 1989b). In the case of home improvement, the intention is that grant-aided repairs and improvements will continue to be available, but the emphasis is to be linked to maximizing the skills and resources of the private sector and the levering of private investment. Indeed, the government's proposals mirrored the conclusions of the Inquiry into British Housing (National Federation of Housing Associations, 1985a, 1985b) that local authorities should not seek to meet all the needs themselves, but should coordinate and facilitate action by other agencies. In 1987 the Minister of Housing outlined the government's stance on private solutions to inner-city housing problems, as to

> [take] forward a range of housing initiatives which will contribute to renewal of inner city housing. These include reform of the improvement grant system to target resources where the needs are greatest; a new power for local authorities to provide assistance for privately rented accommodation; deregulation of new private sector lettings so as to bring new investment into housing associations; and housing action trusts which will take responsibility for run-down housing in designated areas, improve it and pass it on to different tenures and ownerships. (Waldegrave, 1987)

The intentions are reiterated in the government's Expenditure Plan 1989-90 to 1991-92 (H.M. Government, 1989a).

Within this "market-based" approach, area improvement policies are to be retained, but HAAs and GIAs are to be swept away and replaced by a single type of statutory renewal area linked to securing private investment. Again, it is appropriate to question the basis on which these proposals have been formulated and, in particular, whether there is empirical evidence to support the validity of the proposed approach. Chambers and Gray (1985) attribute this model for the revised area improvement approach to the pilot scheme, launched in 1982 in Hammersmith and Fulham, of a Housing Improvement Zone (HIZ). Indeed, the then Minister for Housing stated:

> Public funds are being used for pump priming but the availability of funds from private institutions—for example £2 million special allocation by the Leeds Permanent Building Society—is fundamental to the plan. I shall follow the progress of Hammersmith and Fulham's HIZ with great interest and suggest other institutions consider its suitability in their area. (Stanley, 1982)

In any event, there is no evidence that government has monitored the HIZ's progress, although recent research has evaluated its impact (Chambers, 1988). The HIZ was based on a number of assumptions. First, that management on an area rehabilitation basis remained the most appropriate method of tackling the problem of older housing. (It is significant that no detailed analysis of the impact of the 20 HAAs that had been implemented by the authority preceded this conclusion.) The second assumption was that the local authority was able to promote the improvement of housing within a defined area by stimulating private sector initiatives and encouraging not only the personal commitment of local residents, but also the support of both local financial institutions and local building professionals. Based on these assumptions, the council proposed three major HIZ objectives. First, the council proposed an analysis of the causes underlying the lack of personal investment in property in the area. Second, and of equal importance, was the development of the council's role as coordinator and mediator, bringing together all those involved in the improvement process, such that the sort of public investment required in HAAs would not be needed. The retention of the *existing community* was implicit in the third objective of ensuring that *local residents*, (especially those on low incomes), could be assisted into home ownership.

Despite the central importance of these three objectives in the formulation of the HIZ policy, none was satisfactorily achieved in subsequent policy implementation. Most glaringly, on the first objective, the issue of lack of personal investment was at no time addressed by the council or researched during the progress of the initiative.

Turning to the second objective, the council's aim of developing its role as coordinator and mediator and increasing private investments was of limited success. The council did establish links with private sector financial institutions (including building societies and banks), securing "special" allocations of funds and developing internal administrative structures to bring together actual and potential owners and financiers. In fact, these new relationships, structures, and monies were underutilized.

Part of the reason was that the declaration of the HIZ in 1982 coincided with a two-year period of unrestricted availability of house renovation grants. This, in turn, restricted the extent to which those persons improving dwellings made demands on the cooperating financial institutions, which, although allocating funds to the initiative, made them available only on normal commercial terms. No reliable data are available for public versus private expenditure in the HIZ. However, it was reported that after 16 months of operation, public sector grants had been approved amounting to £921,000, with a further £500,000 worth of grant applications received. Over the same period, the cooperating financial institutions lent some £188,000, a public-to-private investment ratio of 7.5 to 1. Significantly, council documentation produced after May 1984 recorded no further financial information. Chambers (1988), in extrapolating these data to the full five-year term, suggests that public investment may have exceeded £5 million, with the lending institutions contributing a total of £400,000 from their special allocation of £4 million.

Evaluation of the third objective of assisting local residents into owner occupation and retaining the existing community is problematic. Data on tenure redistribution show a decrease in private rental of 406 dwellings in five years and an increase in owner occupation of 317 dwellings. Tenure redistribution, of course, may or may not benefit local residents. Council monitoring of movements of local residents from the private rented sector to owner occupation was inadequate. Only 15 sitting tenants bought from their landlords using the funds set aside by the financial institutions for lending to marginal borrowers. It could be argued, of course, that the new owner-occupiers were existing residents of the area who had purchased through the conventional route, but such a claim would need supporting data that are not available. Qualitative research strongly indicates that contemporary local housing market processes, including rapid house price rises, and gentrification acted to exclude low-income local residents (Chambers, 1988).

To summarize the above points, the evidence to emerge from the HIZ experience, despite the caveats, suggests that house rehabilitation is still dependent on public sector finance and that private sector finance is unlikely to offer a solution to inner-city housing decline and specifically the problems experienced by low-income groups. By the same token, the concept of local communities has remained as ill defined in this experiment as in the HAA programs that have dominated the urban renewal scene since 1974.

CONCLUSIONS

It is not possible to draw definitive general conclusions from individual case studies, although the time-series data presented here are a substantial improvement on the snapshot evaluations on which government housing rehabilitation policy appears to be based. No convincing data exist to support the assumption, bound up with area improvement policies, that HAAs benefit the most disadvantaged local residents. HAAs and similar area-based policies give added momentum to contemporary market forces, through further facilitating tenure redistribution and, in particular, the decline of private renting and the expansion of owner occupation.

The present Conservative government is radically remolding area rehabilitation policies. The material presented in this chapter suggests that these proposals are unlikely to be successful, at least as far as the improvement of the housing conditions of disadvantaged households is concerned. For example, in this context there is no concrete evidence that public sector finance can or will be displaced by private investment. Rehabilitation policies and practices need to be set within the broader context of government philosophy and activity as well as national and regional housing market processes. For the 1980s these have resulted in significant tenure shifts—the continued long-term decline of the private rented sector, the relegation of a contracting public housing sector into a stigmatized welfare role, and a marked expansion of home ownership (Ball et al., 1989; Forrest & Murie, 1988). A closely associated development has been an evolving socio-economic polarization between, and in some cases within, tenures. Many of those excluded from owner occupation are both surviving on low incomes and living in inadequate dwellings. It would be wrong to assume, however, that owner occupation provides the panacea for Britain's housing ills. The tenure is not homogeneous, and a significant minority of home owners suffer considerable housing disadvantage in both physical and social terms (H.M. Government, 1988; Karn, Kemeny, & Williams, 1986). The discussion provided in this chapter has indicated that Thatcherism's stance toward rehabilitation has increasingly followed and helped to accentuate these general trends.

Our own view is somewhat contrary to the direction being followed by the present British government. We believe that area rehabilitation policies, as they emerged in the mid- to late 1970s, despite inadequacies in their implementation, form the basis for a socially just approach, but only if combined with policy mechanisms ensuring that the households most in need of improved housing are not simply targeted in legislation but effectively reached in practice. Such mechanisms might vary through municipalization at one end of the spectrum to socially sensitive approaches to the allocation of public and private monies for home ownership at the other.

NOTES

1. For a full discussion and supporting evidence of this point, see Chambers (1988).
2. The normal period of operation of an HAA is five years, extendable for a further two years.

REFERENCES

Ball, M., Gray, F., & McDowell, L. (1989). *The transformation of Britain*. London: Fontana.

Brindley, T., & Stoker, G. (1988). Housing renewal policy in the 1980's: The scope and limitations of privatization. *Local Government Studies, 14*(5), 45-67.

Cameron, S., & Smart, R. (1982). Area improvement in Newcastle: Past and present. *Housing Review, 31*(6), 216-218.

Central Statistical Office. (1988). *Social Trends 18*. London: HMSO.

Central Statistical Office. (1989). *Annual Abstract of Statistics* (No. 125). London: HMSO.

Chambers, D. (1988). *Discretionary housing policies in three inner London boroughs.* Unpublished doctoral thesis, University of Sussex.

Chambers, D., & Gray, F. (1985). Housing improvement zones: A right move for area improvement? *Housing Review, 32*(3), 90-92.

Channon, P. (1974). *Hansard, 868,* column 1063-1064.

Cockburn, C. (1977). *The local state: Management of cities and people.* London: Pluto.

Crook, A. D. H. (1986). Privatisation of housing and the impact of the Conservative government's initiatives on low-cost home ownership and private renting between 1979 and 1984 in England and Wales: 1. The privatisation policies. *Environment and Planning A, 18,* 639-659.

Cullingworth, J. B. (1979). *Essays on housing policy: The British scene.* London: George Allen & Unwin.

Doling, J., & Stafford, B. (1989). *Homeownership: The diversity of experience.* Aldershot: Gower.

Doling, J., & Thomas, A. (1982). Disrepair in the national housing stock. *Town Planning Review, 53,* 241-256.

English, J., Madigan, R., & Norman, P. (1978). *Slum clearance: The social and administrative context in England and Wales.* London: Croom Helm.

Forrest, R., & Murie, A. (1988). *Selling the welfare state: The privatisation of public housing.* London: Routledge.

Gibson, M. S., & Langstaff, M. J. (1982). *An introduction to urban renewal.* London: Hutchinson.

Gray, F., & Chambers, D. (1985). A red light for the green paper. *Housing, 21*(9), 16-19.

Greenwood, A. (1969). *Hansard, 777,* column 963.

Henney, A. (1976). The Haringey housing action area project. *Housing Review, 24*(2), 57-60.

H.M. Government. (1973a). *Better homes: The next priorities* (Cm 5339). London: HMSO.

H.M. Government. (1973b). *Widening the choices. The next steps in housing* (Cmnd 5280). London: HMSO.

H.M. Government. (1976). *Housing action areas: A detailed examination of declaration reports* (Department of the Environment Improvement Research Note 2-76). London: HMSO.

H.M. Government. (1978). *An analysis of annual progress reports for 1977* (Department of the Environment Improvement Research Note 5-78). London: HMSO.

H.M. Government. (1985). *Home improvement: A new approach* (Cm 9513). London: HMSO.

H.M. Government. (1987). *Housing: The government's proposals* (Cm 214). London: HMSO.

H.M. Government. (1988). *English house condition survey 1986.* London: HMSO.

H.M. Government. (1989a). *The government's expenditure plans 1989-90 to 1991-92* (Cm 609). London: HMSO.

H.M. Government. (1989b). *Local government and housing bill.* London: HMSO.

Karn, V., Kemeny, J., & Williams, P. (1986). *Home ownership in the inner city: Salvation or despair?* Aldershot: Gower.

Kintrea, K. (1987). *Arresting decay in owner occupied housing? The Neighbourhood Revitalisation Services Scheme: A preliminary analysis* (Discussion Paper 13). Glasgow: University of Glasgow, Centre for Housing Research.

Murie, A., & Malpass, P. (1987). *Housing policy and practice*. London: Macmillan.

National Federation of Housing Associations. (1985a). *Inquiry into British housing: The evidence*. London: Author.

National Federation of Housing Associations. (1985b). *Inquiry into British housing: Report*. London: Author.

Ormandy, D. (1985). Implications of the green paper. *Roof, 10*, 4:6.

Roberts, J. T. (1976). *General improvement areas*. Farnborough, Hants, England: Saxon House.

Rowland, A. (1984). Parfett Street housing action area: Improvement of some of London's worst housing. *Housing Review, 33*(3), 82-84.

Spencer, K. M. (1970). Older urban areas and housing improvement policies. *Town Planning Review, 41*, 250-262.

Stanley, J. (1982, June 11). [Address presented at the Annual Conference of British Housing, Brighton, England].

Thomas, A. (1985). *House improvement and agency services: The experience of the Tiverton Road Home Improvement Agency* (Research Memorandum 105). Birmingham: University of Birmingham, Centre for Urban and Regional Studies.

Thomas, A. (1986). *Housing and urban renewal: Residential decay and revitalisation in the private sector*. London: Allen & Unwin.

Turney, J. (1980). Success for the first housing action area declared: But is it working? *Housing Review, 29*(5), 142-145.

Waldegrave, W. (1987). *Hansard, 11*, column 525-526.

Young, M., & Willmott, P. (1957). *Family and kinship in East London*. London: Routledge.

Part VI

Developments in Socialist Economies

Introduction

IVÁN SZELÉNYI

THE FOLLOWING TWO case studies report on the paralysis and what appears to be the inevitable eventual collapse of the socialist housing economy, which coincides with a similar paralysis and collapse of the national economies of these countries. Both the Hungarian and the Yugoslav cases show us that attempts to "fine-tune" the socialist housing economy without confronting and trying to find solutions for the fundamental contradictions of this housing model have proved to be only half reforms or, worse, quasi-reforms, and they are doomed to failure. This introduction puts the evidence presented in these chapters in a broader theoretic perspective.

What are the fundamental contradictions of socialist housing that must be attacked by effective, genuine reform? Hegedüs and Tosics accurately define key problem of socialist housing: This is an "economic model in which wages . . . [do] not contain the costs of housing and those of infrastructure in general."

THE PROBLEM OF SOCIALIST HOUSING SYSTEMS AS A PROBLEM OF PROPERTY RIGHTS

I use property rights and principal-agent theory to situate my argument theoretically. Let me do this first concerning the socialist economy in general, and then specify it for socialist housing systems.

Socialist reforms until recently faltered because they tried either decentralization of decision making or simulation of markets without altering the bundle of property rights. This proved to be quasi-reform. Over the last three or four years we have begun to understand that the tendency of socialist economies to produce systematic shortages and their inability to adapt in a flexible manner to the challenges of world markets are related to certain restrictions of property rights, to a certain "misspecification" of the bundle of property rights. Thus successful reform will have to move beyond reform of economic mechanisms; it has to be a reform of property relations.

A similar argument can be made about housing reform. The lack of dynamics and unjust allocation in socialist housing systems may be attributable to the misspecification of property rights. More dynamism and more socially just allocation can be achieved only through changes in those rights, rather than simply through changing the levels of administrative decision making (for instance, by decentralization) or by simulating markets either within the public sector or outside the public sector, but without the creation of property rights that are consistent with the logic of market integration. There are at least two such misspecifications of property rights in connection with the housing system.

(1) Restrictions of the property rights of wage earners. The problem that is defined as the key one by Hegedüs and Tosics—namely, the lack of accumulation of housing costs in incomes—can be interpreted as a problem of property rights. The government or the publicly owned firm retains the right to dispose of the component of the salary that is supposed to cover costs of housing. In doing so, it restricts the property rights of the workers, or the "owners of labor power": It does not allow workers to make some of the fundamental decisions about the "reproduction" of their labor power. The history of socialist housing indicates that public firms and governments have been agents who misrepresented their principals, employees, or the whole of the citizenry in more than one respect. First, as both chapters show, government planners or public firm managers over the long run were likely to invest less in housing than wage earners would have if they had been in the position to make these decisions. Second, government planners, administrators, and public firm managers systematically allocated housing on the grounds of merit to the better educated, people with higher incomes. Thus the overall result of this "misrepresentation" has been relatively little housing construction and socially unjust allocation.

(2) Restrictions of the property rights of real estate owners and developers. In all socialist countries, ownership of private housing has been restricted to one or two units. Construction or management of

housing for profit has not been allowed. As a result, no legalized private rental sector developed and personal savings were not mobilized toward the real estate sector. This proved to be devastating. In Western market economies, middle-income managers, professionals, academics, and small business owners often put some of their savings, as a form of pension insurance, in the real estate market. By not allowing private ownership of housing for profit, socialist planners aggravated the tendency to accumulate less and to overconsume, reduced the flow of capital into the undercapitalized housing sector, restricted petty capital accumulation, and may even have contributed to the excessive inflation of housing on the thus restricted housing market.

Without addressing this issue, without changing these property relations, neither proper labor markets nor proper housing markets are possible, and reforms will remain half reforms. I will illustrate the relevance of these theoretical propositions with examples from the two chapters.

THE FAILURE OF YUGOSLAV HOUSING POLICY: WHY DECENTRALIZATION AND SELF-MANAGEMENT ARE NOT ENOUGH

In Yugoslavia, the story begins with *decentralization,* first from the central state to the local state (during the early 1950s), then from the government to the publicly owned but self-managed enterprise (this move began in 1965 and accelerated in 1974). None of these strategies worked. Enterprises and even municipalities allocated housing according to merit, rather than need, thus leaving low-income groups without housing. This redistributive injustice was coupled with inefficiency: Enterprises had little vested interest in building much housing anyway. In the language of principal-agent theory, the enterprises proved to be poor agents of their employees in satisfying their housing needs. The principal-agent dilemma in this respect is as acute or even more acute between enterprises and employees than it is between state and citizens. In other words, attempts to decentralize decision making without altering property rights avoid and probably aggravate, rather than solve, the problem. As similar half reforms paralyzed other sectors of the economy, by the 1980s the volume of housing construction began to fall. An increasing proportion of the population was left without an "agent" and with no income to compete on real markets. Thus it had no choice but to try to solve its housing problems outside the established economic institutions, in the "informal economy."

THE FAILURE OF
HUNGARIAN HOUSING POLICY:
WHY SIMULATED MARKETS DO NOT WORK

The Hungarian story begins differently but ends similarly. Instead of decentralizing, economic reformers tried to use the *market*. But the attempt to commodify without allowing private ownership culminated in a crisis that is almost as grave as the Yugoslav one. Since neither the established quasi-markets nor the administrative sector works, many Hungarians are left similarly dependent on the "second economy."

First the Hungarian planners began to allow market forces to operate outside the state sector (this began during the 1960s, then accelerated in 1971). *Market* in the first instance meant that the government allowed people to build their own homes from their own savings and/or with their own labor. But this at no point and in no way implied the creation of a real private sector or real market. Trading/building homes for profit was not allowed, let alone encouraged. Further price reforms and changes in allocation mechanisms were never followed up with wage reforms. Still, despite these limitations, the opening up of markets proved to be a success during the 1970s. Hungary achieved reasonable levels of housing construction. The emergent new market sector of housing began to attract some of the higher-income groups away from the state sector, thus enabling the state sector, which for a few years retained high levels of public housing construction, to begin—at least partially—to perform some welfare functions. One would have hoped, by the mid-1970s, that the reformers would stick to this approach, thus continuing the deregulation of private activities while maintaining the levels of public housing needed until housing shortages could be eliminated. Instead, by the early 1980s Hungarian reformers began to commit a double error: (a) They kept the controls on private activities, but (b) they began to scale back public expenditures, to commodify the public sector, increase rents, sell public housing, and reprivatize the public sector, without addressing the misspecification of property rights. There was not enough deregulation for the private sector and too much reprivatization and commodification in the public sector. This proved to be a recipe for disaster: The volume of new construction declined and the poor, unable to compete in the market, remained unable to find accommodation in the shrinking public sector.

Thus the Hungarian and Yugoslav case studies indicate that the housing reforms, like economic reforms in general, did not overcome the failures of socialist redistributive economies. Instead of diversify-

ing property relations, they tried to solve economic problems by rede-
fining the levels at which decisions are made or simulating markets
within the public, redistributive sector.

WHERE TO GO FROM HERE?

My impression is that housing policy experts in Eastern Europe have
run out of ideas—they just do not know what to do. In the conclusion
to their chapter, Hegedüs and Tosics acknowledge that while the "original model began to give way to another housing system with a completely different logic . . . [it] is still uncertain which type of housing
system will finally replace the original housing model." They consider
two alternatives: (a) the *social democratic model,* in which "the role of
the state is reduced to handle social needs"; and (b) a system in which
the state withdraws from housing directly, but retains some influence
over the market process. The first model means that the state retains a
sector of public housing for those in need; in the second case no
public housing is retained—the state interferes in market processes
only through tax concessions, interest rate subsidies, and the like. Thus
Eastern Europe will look like Sweden or will look more like the United
States.

While I think drastic deregulation for private housing is inevitable,
I would be less sanguine about the possibility or even the desirability
of elimination or drastic reduction of the sphere of public housing. I
have identified two sets of property rights problems in connection with
the socialist housing economy. It is easy to deal with the second one, to
allow private housing—no major risks there. Such a private sector
promises to be small anyway: Demand for such housing is limited. But
it is formidably difficult to do much about the first set of property rights
problems: to conduct wage reform that reallocates costs of housing into
wages and salaries. These resources have disappeared from the government budgets; there is not much to reallocate into incomes. A massive
increase in incomes seems also to be impossible in the public sector,
which could not absorb any increase in its costs of production. The
whole sector barely functions with its current level of expenditures.

Thus I anticipate the change to be gradual and slow. My forecast is
that the public sector will remain large in the sphere of production for
at least another decade. It employs a labor force of which a significant
proportion is unemployable elsewhere. Without risking massive unemployment and political instability inevitably caused by such levels of
unemployment, these public firms should be kept in business even if on
the whole they just cover costs and do not generate profits comparable

to those made in the private sector. I would not be surprised if 10 or 15 years from now the public sector still employs 40-60% of the work force and still pays them wages well below those paid in the emergent new private sector. This would mean that half of the population will depend on public housing. Under such circumstances, the last thing I would recommend to Hungarian or Yugoslav public authorities is to cut back on public housing. On the contrary: I would recommend that they move back to those levels of public expenditures on housing that existed during the early or mid-1970s. We should hope that the next decade, after more than 50 years of stagnation caused by the war and communism, will be finally an epoch of dynamic growth in housing. I would advocate strongly against further reprivatization of public housing and cutbacks in public expenditures, and I would recommend increases in public expenditure and strong support for profit-oriented construction by private investors.

If Eastern Europe opts for massive reprivatization in the spheres of both production and housing, and for the so-called reprivatization shock therapy recommended by several Hungarian economists, then we can anticipate very high (25-30%) unemployment for at least one or two decades. While some workers will begin to earn significantly higher incomes than they earn today, many will fall into poverty and will not be able to provide housing for their families. Such an economic strategy is very possible and—as the massive sale of Hungarian public property to Western firms indicates—already very much under way. It would push Eastern Europe from stagnating state socialism straight into dependent capitalism. The results of such development are well known—and devastating.

14

Moving Away from the Socialist Housing Model: The Changing Role of Filtering in Hungarian Housing

JÓZSEF HEGEDÜS
IVÁN TOSICS

BY THE MID-1980s, the total volume of house building in Hungary dropped to about half to two-thirds of the figure 10 years earlier. The rate of state (budget-financed) house building plummeted from the 35% characterizing the previous three decades to 10%. These signals show deep structural changes in Hungarian housing.

One of the most important theoretical papers dealing with Eastern European housing is that of Iván Szelényi (1983), who states that state distribution is the dominant integrating mechanism in socialist housing, and that the inequalities in housing arise from this mechanism. The market is a secondary mechanism, the role of which is to compensate for the inequalities caused by the state. This theory is Szelényi's attempt to describe the 1960s in Eastern European housing.

The 1971 Housing Act (housing reform) changed the roles of institutions in building and allocating housing, and this seemed to be a move toward the market system. Based on this new development, Szelényi modified his theory, claiming that since the 1970s the role of market distribution had been changed (Szelényi & Manchin, 1987). As the importance of the market grows in housing, it contributes increasingly to inequalities. In this view, the basic model has not changed: The dominance of state distribution holds on. The new feature of this model is a secondary mechanism that also works to produce inequalities.

Our research has not confirmed entirely the above-outlined hypothesis. We have critical observations on two points. First, in the 1970s, state distribution, not the market mechanism, strengthened, and the social policy functions of state distribution became more marked (Lowe & Tosics, 1988; Tosics, 1987). The advance of the market started only in the 1980s. Pickvance (1988) interprets the developments from the middle of the 1970s as a growing influence and regressive intervention of big enterprises dominating socialist industry. Second, we cannot accept the interpretation of the role of the state and of the market in Szelényi's theory. In our opinion, whether state intervention has regressive or progressive effects depends on particular social and political conditions. Besides this, in the private sphere, it is not the market but the self-help form that has resulted in a more equitable allocation system (Hegedüs, 1987a).

The Housing Act of 1983 has been a decisive step toward a more market-type housing system, but even in this "transition period" state and market (and self-help) distribution play peculiar roles that cannot be handled by the theory outlined above. The really important results emerge as a consequence of the interaction of these distribution mechanisms.

In the first part of this chapter we summarize the main points referring to the operation and inner conflicts of the Eastern European housing model. In the second part, based on empirical data, we seek to identify the crucial elements in the social process leading to the change of the model in the transitional period.

A SOCIOLOGICAL MODEL OF THE EASTERN EUROPEAN HOUSING SYSTEM[1]

THE MODEL AND THE CRACKS

The common source of Eastern European housing systems was an economic model in which wages did not contain the costs of housing and those of infrastructure in general. As expressed officially, the state had to provide for housing (and other infrastructural items, such as education, health care, and transportation) through central redistribution of the national income. The policy, aimed at forced economic growth and industrialization, concentrated investments mainly in urban regions and some rural areas with industrial background (chiefly mining). Accordingly, housing needs were supposed to appear in these regions, and the housing-provision role of the state was restricted to these more developed areas. In underdeveloped rural areas the private sector (i.e., self-help building) was to solve the housing problem. In short, in the socialist housing model the population had no right and no possibility to decide on the level and means of housing consumption; there was no (or only very indirect) feedback between the production and the consumption of housing.

Even if this model had not existed in its pure form, it serves as a starting or reference point from which the line of actual developments of housing policies in Eastern European countries from the late 1940s can be interpreted. The conditions for the operation of this model were strict: The state had to invest on a large scale in urban housing; the influx into cities had to be controlled; the distribution processes had to be dominated by state allocation.

In Hungary—as in other Eastern European countries—infrastructural investment remained very low among the economic developmental priorities because of the policy of forced economic growth. This policy minimizes the state housing supply. On the other hand, there has been, from the very beginning, a continuous dynamic growth in demand for urban housing because of forced industrialization, which has led to increasing urban housing shortages (Konrád & Szelényi, 1977).

Thus the socialist housing model implied tensions from the beginning, as even the minimal conditions for its operation were lacking. After exhausting the infrastructural reserves, "cracks" emerged on both the supply and the demand sides.[2]

The cracks on the demand side have to do with additional purchasing power that does not fit into the logic of the original model:

(1) The Eastern European economies are economies of shortage (Kornai, 1983), where "forced saving" is a systematic element of the economic system. Basically, this means an excess demand that materializes to a significant extent in demand for housing.

(2) A thin high-income layer appears in the state sector (artists, physicians, and so on) who are willing to spend their excess income on housing.

(3) Through the second economy[3] (spreading gradually since the economic reform of 1968), uncontrolled income rises and uncontrolled secondary income distribution takes place; one of the main targets of this excess income is housing.

(4) The secondhand marketing of state housing (tolerated commodification of state subsidies) has become the source of excess demand for private house building.

Cracks also appear on the supply side, indicating that the state cannot fully control the housing supply:

(1) As a result of demographic processes (deaths, marriages, and so on), homes are vacated and marketed. Even if the number is small, they create a real second market.

(2) The state building industry produces housing for the "market" (financed by the National Savings Bank, or NSB) on a limited scale that does not mean a real market mechanism, only a selling policy closer to the real value of the commodity.

(3) Private self-build housing emerges on urban housing markets on a small scale in spite of the state's attempt to wipe it out.

(4) A private market of house building is strengthened that is based on the second economy, where the workers of the first economy are employed in second shifts and on weekends.

In principle, these divergences from the starting model could be controlled by the state. On both the supply and the demand sides these cracks could be filled, these dysfunctions could be localized and stabilized, through strict income and labor policy measures. Obviously, this type of administrative regulation always has a price. The restrictive policy affects above all the second economy, because this is the main reason for the widening of the cracks. The gravest consequence could be a loss of efficiency in the second economy, which in turn may result in economic difficulties. Therefore, Hungarian policy preferred compromise solutions over strict controls on demand and supply. This strategy—unique in Eastern Europe until recently—contributed to the gradual changing of the socialist housing model.

DISTRIBUTION MECHANISMS IN THE STATE AND THE PRIVATE SECTOR

The basic conflict of the socialist housing model is between the state and the private sector. The condition for the survival of the model is the dominance of the state housing sector in urban areas, but its form and range could change. In the early period of socialist housing policy, the institutions with a role in the housing market made definite efforts to control the whole housing market (Hegedüs & Tosics, 1983, pp. 475-477). Their endeavor had clear limits. The economic costs of practicing total direct control over production and distribution processes were too high. The role of the state was changing depending on these factors (cost of control, willingness to control the whole process). This, however, did not affect the dominance of the state, which as a minimum condition meant the disposition over the highest-value housing and the control of the resources distributed through the budget to housing, giving systematic advantages to privileged social groups. State dominance is not the same as state control over the housing market: State dominance can be upheld at different degrees of state control.

State intervention in the housing market does not take place according to homogeneous distribution mechanisms (Hegedüs & Tosics, 1988). We can differentiate two characteristic distribution principles in the state sector:

(1) the *social* (welfare) principle, by which the tenant or buyer of the flat is chosen from among people in needy situations, the social effect of which is a compensation for inequalities in housing

(2) the *positional* principle, by which particular (not legalized) interests of the distributor determine the allocation—that is, applicants' access to housing is influenced by their social status and position (Szelényi, 1983)

Also, the transactions in the private sphere take place according to different distribution mechanisms:

(1) the *reciprocative* principle in self-help housing, by which the market forces are substituted or complemented by reciprocative social connections, having little to do with social position (see Hegedüs, 1988; Sik, 1988)

(2) the *market* principle, which contains the second market of used homes and the market of private building

The latter means building in high-value urban areas (inner city, green belt). Access to these new homes is determined by social status, wealth, and income. The second market of used homes predominantly denotes the buying and selling of the lower-value part of the housing stock.

Our research has shown that different combinations of the above-outlined distribution mechanisms can be consistent with the socialist housing model.

STRUCTURAL CHANGES IN THE HOUSING MARKET

The basic dilemma of socialist housing policy is that the state cannot solve the urban housing problem exclusively from budgetary resources. It needs the private sector. But the involvement of the private sector (provided that it extends to the high-value urban areas) entails the risk of decreasing state dominance. Moreover, it puts an extra burden on the population whose source of extra income is the second economy, as wages in the first economy do not contain the costs of housing. The upholding of the socialist housing model depends on whether the second economy and the private sphere can be involved in housing without limiting state dominance.

In the original Eastern European housing model the dominance of state house building means that in areas with the best infrastructure above-average homes are built, while private house building is pushed out of the infrastructurally higher-quality urban areas and produces homes far inferior to state housing. According to our hypothesis, up to the first part of the 1960s the dominance of state housing was not questioned in the Hungarian housing system: The highest-quality houses built in high-value urban areas (new housing estates near the city centers) were controlled by state institutions.

From the 1960s, private house building started to change. Housing became the most important aim of family savings, and incomes from the second economy (in the beginning affecting mainly rural areas) were spent on housing.[4] This meant that better-quality homes were built in the private sector. In the state sector, factors limiting the quality of the newly built flats grew in importance; the quality, size, and location of homes did not improve in order to increase the number of units built.

In this period the relative quality of state housing deteriorated, but its "ecological location" remained much better than that of units in the private sector. The dominance of the state sector was not entirely questioned in this situation because private house building was not allowed in higher-value urban areas (these had been reserved for future state house building). The secondhand market contributed to the dominance of the state indirectly as the products of the state industry were overvalued because of the urban housing shortage.

The validity of the basic model began to be questioned only when private house building was allowed to move toward the best-quality, highest-value urban areas, and when—partly because of this shift—the secondhand market revalued the relative position of the different housing classes. In our opinion the collapse of the original model (that is, the vanishing of state dominance) occurs when the best houses are built by the private sector and are allocated without any direct state intervention.

SOCIAL CONFLICTS
OF THE TRANSITIONAL PERIOD

CONTROLLED OR MARKET FILTERING

Up to the late 1970s, the logic of the socialist housing model did not change essentially despite some significant changes in housing policy. The dominance of the state was not jeopardized by the cracks concerning the demand and supply of housing, by modifications of the allocation principles, or by alterations in the structure of the housing market. Substantial changes in the basic model occurred only at the end of the 1970s, when the economic conditions radically deteriorated. During the early 1980s the state was compelled to change its housing policy completely, to give up its dominance, and to acknowledge (even support actively) the growing role of the market. We call this period of change a *transitional period*.

The problems of the model became more and more intense even in the 1970s, in a period of relatively favorable economic conditions and increasing state housing investments. On the one hand, regarding the demand side, the differences in population incomes increased as a result of the emergence of the second economy after the 1968 economic reform. Those social groups that could acquire high incomes from the second economy were no longer satisfied with state flats and wanted (and started) to build homes in better parts of cities. These prestigious housing areas (most of which had been reserved for future state housing) kept shrinking, and prices started to increase rapidly. These housing market processes were really governed by market principles, and in these "best" housing submarkets high inflation rates were developed. At first glance this might suggest that the weight of the market mechanisms increased very significantly within the whole housing system (an interpretation accepted by Szelényi & Manchin, 1987). In the course of the 1970s, however, these changes were very limited in volume.

On the other hand, regarding the supply side, the quality of state housing (size of flats and so on) was highly constrained and did not change for years, in keeping with the aim of eliminating quantitative shortages. Dissatisfaction with these average typical flats increased even among families who got them with high subventions. The position of state housing was further weakened by the appearance of outskirt high-rise housing estates having higher and higher construction costs but less and less financial gain for the population (because the market value, the basis for the resale price of these flats, was not increasing very fast). In contrast, private house building gained better position in the housing hierarchy: its average flat size rose from 71 m^2 (1975) to 85 m^2 (1981), while the same figures in state housing stagnated (52 m^2 against 54 m^2).

A central question at the end of the 1970s was how to satisfy simultaneously the still substantial quantity needs and growing quality requirements. According to this terminology, the quantity-type need is tied to "homeless" families (new households, living in sublet or shared homes with parents), while the quality-type need is tied to families with rented or owned homes aspiring to move up in the market. From the late 1970s the state housing policy tried to maintain its dominance over the housing market with the strategy we have termed *controlled filtering*. The point of this housing policy practice was that the state (the local councils) tried to satisfy the quality-type housing needs within the model of state housing in a way that was suitable to tackle—with the vacated flats—the quantity-type needs as well. Families with quality need got access to new (rented or privately owned) dwellings, and their previous dwellings were reallocated by the local council. As a sign of this strategy, from the late 1970s families with quality need had priorities on waiting lists.

The housing policy containing controlled filtering could not hold its ground for long. The experiences of the 1971 Housing Act have shown that the increasing role of the private sector[5] restricted the control of the state over the housing market, but at the same time it contributed to fulfilling the housing policy tasks the state had taken on. This policy worked quite well until the end of the 1970s, when the debt-related economic crisis broke out. The gap between housing prices and first economy incomes was growing rapidly; the state was no longer able to finance the increasing subsidies to the state building industry and had to cut back the supply of state-built flats.

The inevitable change in housing policy occurred in 1983, when a new Housing Act reevaluated the connection between the state and the private sector. The aim (similarly to 1971) was to release state control over the private sector in order to decrease financial pressure on the state budget. In 1983 nearly all forms of "deep subventions" (Hegedüs & Tosics, 1983) connected to state-built housing were abolished and spread toward privately built homes. Terms of loans were improved and housing allowances were available for private builders; also, the tax discrimination toward private market transactions was ended. Thus the Housing Act wiped out the discrimination of the private sector.

The new way of satisfying quality needs became state-supported "market filtering," in which families aspiring for better flats satisfy their needs with state subsidies, but outside the state sphere. This means that the state gave up its direct control over these mobility channels (i.e., over the best housing classes). In the new model most state subventions are allotted to families moving from the state sector to new private housing. To promote this type of mobility between spheres, the state gives more possibilities for private building in areas with better infrastructure (lifting building prohibitions, creating and selling building plots in better parts of cities, selling out parts of building land already prepared and reserved for state building).

STRUCTURAL CHANGES IN
THE LOCAL HOUSING MARKET:
THE SHIFT TOWARD MARKET FILTERING

In the first part of the 1980s, we carried out empirical housing market research in three county seats, applying, among other methods, that of the vacancy chain studies. In these surveys we measured how many vacancies had been created with the different forms of new construction (that is, the average lengths of the chains), which gave us an indicator of the filtering processes as well. (For detailed results and methods, see Hegedüs & Tosics, 1988.) The average length of chains was 1.44 for new homes built in the period 1980-82. Vacancy chain studies carried out in other countries usually report 1.7-2.5 average length of chains, which is significantly higher than our figure even if we are fully aware of the dangers in direct comparison (see Sharpe, 1978). First-time buyers, renters, and builders are more typical actors in the whole housing market, and their problems are more decisive in the allocation processes than those of families who already have flats. In spite of all these facts, it is worthwhile to investigate the filtering process in detail because this makes it possible to highlight some important aspects of the recent changes in the Hungarian housing system.

After January 1983 (the introduction of the new Housing Act) until 1985, there was only a slight increase in the average length of chains, from 1.44 to 1.53. The new housing policy had a more marked impact on the relative rate of the state and private sector as the share of the state sector dropped from 79% to 57% in house building.

Table 14.1 shows the composition of filtering processes by mobility types. Before 1983, controlled filtering had an overwhelming weight: Chains starting from the state sphere amounted to 79% (71% + 8%) of total mobility. In comparison, the rate of market filtering at 21% (11% + 10%) was very low. From 1983 onward the weight of controlled

TABLE 14.1

The Structure of the Mobility Processes

(Moves of Families Already Having Flats) (in thousands)

Transition Matrix			*To*			
		1980-1982			*1983-1985*	
From	*State Sphere*	*Private Sphere*	*Total (N = 277)*	*State Sphere*	*Private Sphere*	*Total (N = 114)*
State sphere	708	113	821	529	358	887
Private sphere	80	99	179	44	69	113
Total	788	212	1000	573	427	1000
Average length of chains	1.43	1.47	1.44	1.50	1.56	1.53

filtering decreased dramatically, even if it is still above 50%, and market filtering represents a comparable magnitude (36% plus 7%).[6]

With the help of vacancy chains it is possible to illustrate the peculiar features of the housing market analyzed in the first part of the chapter. The quality of flats in state house building—measured by the number of rooms, floor area, and "value" of home[7]—is far below that of homes in private house building. Table 14.2 shows that the difference between the two categories already before 1983 was almost one room, around 50-60 m² and about 250,000-300,000 forints, and these differences grew after 1983. Because of the relatively small difference in chain length, it would be possible to say that the state sphere is able to generate the same mobility with much lower-quality homes. This, however, does not mean that the state sphere is more efficient, because efficiency depends on the costs invested in generating this mobility.

In Hungarian cities the inner parts are usually much more valuable than the outer parts or the areas around the cities. From Table 14.3 it can be seen that in the first period the better parts of the cities were dominated by the state sphere. After 1983, however, the locational advantages were shifted from the state to the private sphere; within the latter the proportion of homes built in central areas increased (from 9% to 57%). Thus privately built urban housing, which had already reached a high level in respect to floor space, number of rooms per house, and the like, became unambiguously the best housing class after 1983.

TABLE 14.2

Average Number of Rooms, Floor Space, and Value of New Flats

| | 1980-1982 | | 1983-1985 | |
	State Sphere	Private Sphere	State Sphere	Private Sphere
Chains with more than one link				
number of rooms	2.44	3.37	2.26	3.36
floor space[a]	64.0	120.1	55.5	110.2
value[b]	753	1012	790	1281
number of cases	196	53	81	61
Chains with only one link				
number of rooms	2.09	3.14	1.90	3.01
floor space[a]	57.1	114.9	51.7	113.6
value[b]	672	811	663	1058
number of cases	411	107	122	103

a. In square meters.
b. In thousands of forints.

TABLE 14.3

Location of New Homes at the Starting Point of the Filtering Process
(Chains with More than One link)

| | 1980-1982 | | 1983-1985 | |
	State Sphere	Private Sphere	State Sphere	Private Sphere
Inner parts of cities	34.2	9.4	47.6	57.4
Outer parts of cities	65.8	62.3	52.4	31.1
Environs of cities	—	28.3	—	11.5
Total	100.0	100.0	100.0	100.0
Number of cases	196	53	81	61

CHANGES IN THE DISTRIBUTION MECHANISMS

The quality of new state-built homes remained at the same level or even decreased slightly between the two periods under investigation. One possible explanation for this is that the pressure on improving quality eased as the social groups having high quality needs left this sector.

TABLE 14.4

Indicators of Social Status of Families Acquiring New Flats

	1980-1982		1983-1985	
	State Sphere	Private Sphere	State Sphere	Private Sphere
Chains with more than one link				
school qualification (years)	12.0	11.3	12.6	13.4
average income per capita (forints)	3296	3131	3821	4529
high-status families (%)	32.6	26.4	37.0	55.7
number of cases	196	53	81	61
Chains with only one link				
school qualification (years)	11.2	10.4	10.9	11.1
average income per capita (forints)	3036	2856	3478	3716
high-status families (%)	19.1	10.9	6.6	15.1
number of cases	411	107	122	103

In the 1980-82 period the social status of the families moving to new state-built homes as the starting links of chains was higher than that of the families moving within the private sphere (Table 14.4). The high-status groups' moves to the best part of the housing stock (mediated by the market in Western housing systems) was managed by the state in Hungary through controlled filtering. After 1983, however, the subsidized market filtering takes back its theoretical role: Higher social groups move through private house building to their new homes. As a secondary consequence of this the social features in state distribution are strengthening, concerning the allocation of flats to homeless families.

Table 14.4 shows how the shift toward market filtering substantially increased the status indices of social groups moving to the starting home of the chain in the private sector. On the other hand, the social status of the families moving to the starting home of the chain in the state sector did not decrease. Controlled filtering maintained its relatively high status, because the state had "reserved subsidies" [8] that made it possible to get a high subsidy within state distribution even after 1983. One form of this was the change in tenure, that is, moving from the rental sector to the owner-occupied sector through state allocation. In this way families could acquire large increases in the values of flats (subsidies) even if they had not substantially improved the quality of

TABLE 14.5

Mobility Processes Within the State Sector

(Moves of Families Already Having Flats)

| | To | | | | | |
| | 1980-1982 | | | 1983-1985 | | |
From	Rented Flat	Owner- Occupied Flat	Total (N = 196)	Rented Flat	Owner- Occupied Flat	Total (N = 60)
Rented Flat	463	165	628	140	291	431
Owner-Occupied Flat		372	372		569	569
Total	463	537	1000	140	860	1000

flats. The other form was changing the location (ecological position): moving from the outskirt housing estates to the inner city.

In the 1980-82 period the majority of moves within the state sphere fell into two categories: moves within the rental sector and moves within the owner-occupied sector (Table 14.5). Mobility between the two sectors was relatively insignificant. After 1983 the inner mobility of the rental sector decreased; the moves from the rental to the owner-occupied sector became more significant and also the share of moves within the owner-occupied sector increased. Financial advantages of tenure change made state-built (owner-occupied) housing desirable for some higher-status families, too.

The change in the character of controlled filtering can also be proved by the ecological position of new housing. In the period 1980-82 the mobility within the state sphere typically meant moves between flats in outer housing estates; after 1983 the share of flats with better (inner-city) ecological position increased. These inner-city, state-built, owner-occupied flats became more attractive even if they were not larger than state-built flats in outer housing estates.

The social consequences of the changes in controlled filtering can be seen in Table 14.6. The social status of families moving from rented flats to other rented flats decreases after 1983, while the social status of those moving from rented to state-built, owner-occupied (NSB) flats increases. Thus the relatively less subsidized moves between state rented flats can be characterized with the social distribution principle, while the more subsidized moves from rental into state-built NSB flats are closer to the positional distribution. The high subsidies given to the

TABLE 14.6

Rate of Higher-Status Families (Managers, Intellectuals, and
White-Collar Pensioners) Within Families Getting New Flats in the State Sphere
(percentage independent in each cell)

| | To | | | |
| | 1980-1982 | | 1983-1985 | |
From	Rented Flat	Owner-Occupied Flat	Rented Flat	Owner-Occupied Flat
Rented flat	24.0	62.0	0	70.8
Owner-occupied flat		32.8	—	30.8

latter form of mobility make state-built NSB flats much cheaper for
tenants than moves from the rental sector to private house building.

CHANGE IN THE TYPE OF FILTERING: ADVANTAGES AND DISADVANTAGES

When evaluating controlled and market filtering, we have to go back
to the basic idea of filtering policy.[9] The well-known dispute over this
idea has been focused on the question of whether indirect social effects
actually exist or whether the emphasis on these effects serves more to
cover up (or mask) the direct subsidizing of better-off families.[10]

Controlled filtering is highly controversial. This policy is socially
unjust, not so much because of the use of the principle of positional
allocation but because large subsidies are granted to already better-off
social groups that otherwise have the best possibilities to improve their
housing situation. Controlled filtering is very advantageous for these
families: They get better flats with minimal cash and a lot of credit. This
is why this housing policy could achieve a relatively big chain length
with relatively small number of new state-built flats.

The new housing policy in effect since 1983 has removed the restrictions on private housing. Moreover, enhanced state support for market
filtering has been observable. New measures were taken to encourage
residents to leave the state sector (families giving back their state flats
got many times the sums they had paid).

The effects of the new state subvention policy cannot be fully
evaluated yet. However, it might be assumed that market filtering, as
a new system of state support for housing, strengthens the regressive
tendencies in housing without automatically ensuring any compensa-

tion (i.e., indirect social effect). There is no institutional guarantee to keep the "oversubsidizing" of higher status groups below a certain limit.

In fact, both filtering models seem to strengthen regressive tendencies.[11] Thus the change of the socialist housing model does not mean the elimination of the regressiveness of the housing policy. The change means a modification of the logic of regressiveness, replacing the positional distribution principle with the market principle (influenced by the state) regarding the allocation of advantages. The open question at the beginning of the 1990s in Hungary is whether this regressive, more or less market-oriented housing policy (which differs fundamentally from the original socialist housing model) can be "legitimated," at least temporarily, by arguments referring to the serious situation of the economy. The other question is whether and how this housing policy can be developed toward a more stable and just social model.

CONCLUSION:

THE ALTERNATIVES OF A NEW MODEL

Discussing the new tendencies of the 1980s,[12] we conclude that the original socialist housing model has begun to give way to another housing system with a completely different logic. But it is still uncertain which type of housing system will finally replace the original socialist housing model.

In theory, there are several possibilities. One of them is creating a housing policy similar to the Western European social democratic model. This would mean that the role of the state is reduced to handling social needs, and market processes predominate (not only private provision but the whole housing system). There are some conscious attempts in this direction. However, implementation of such a model has strict political and economic conditions. To fulfill these prerequisites is problematic not only because of the recent economic crisis but also because it would require substantial structural changes in the whole institutional and decision-making system of housing policy and in the allocation of resources in general.

However, modification of the traditional socialist model in another direction is also possible. The withdrawal of the state from the housing sector is not necessarily tantamount to the state giving up all of its possibilities of intervention. The change may lead to an increase in state influence over the market processes (consistent with the Western model). When, however, this "market-regulating role" of the state means the allocation of advantages and disadvantages without social or political control, the positional advantages within state distribution will turn into market advantages. This possibility may come up especially if the gap between the first and second economies keeps widening. In this case the well-positioned social groups within the state sphere will have no way to make good use of their positional advantages other than to convert them into market advantages. The withdrawal of the state from the housing sector may also mean the state's giving up its social functions, which might lead to the spread of a lower-standard variant of already existing self-help housing in rural areas and its expansion to urban regions as well.

It is not yet clear whether the new model in the making will reinforce the market or corruption in the upper segment of the housing system and social policy or squatters' slums in the lower.

NOTES

1. This part of the chapter further develops the ideas presented in three earlier papers (Hegedüs, 1987a, 1987b, 1988).

2. The terms *housing market, demand,* and *supply* are not used in this chapter in strict economic sense, because in the socialist economic system no self-regulating mechanisms are working.

3. The *first economy* is basically the state sphere in Hungary, which has been, up to the present, the dominant actor in the economy. The *second economy* includes economic activities outside the state sphere that, however, maybe be closely connected to state organizations. Fully licensed individual craft workers and artisans, small farmers, as well as illegal entrepreneurs all can be put under the heading of second economy. The term is used here in a broader and slightly different sense than the term *informal economy.*

4. One of the main reasons for this was that there were no capital investment possibilities allowed.

5. The increasing role of the private sector meant that state housing policy was based on families' savings or had promoted self-help building based on second economy activities. All this did not necessarily include more market activity.

6. On the basis of data for 1985-87, when state house building decreased significantly, we can estimate that the share of market filtering grew above 50%.

7. The value of homes was estimated by specialists independent of tenureship, so this figure expresses the use value rather than the exchange value of the homes. The latter two figures differ only in the case of rented homes.

8. We term those financial and allocational methods *reserved subsidies* that enable the moving families to get advantages without causing direct excess expenditures for the budget.

9. For a good summary, see Clark (1984).

10. For the exact formulation of the latter, antifiltering viewpoint, see Boddy and Gray (1979).

11. According to our data, this is not the case with the mobility within the rental sector. However, this subtype of controlled filtering is very limited in volume.

12. It is not possible in this chapter to deal with the latest developments: A new regulation was introduced at the beginning of 1989, and new political developments changed the housing system significantly.

REFERENCES

Boddy, M., & Gray, F. (1979). Filtering theory, housing policy and the legitimation of inequality. *Policy and Politics, 7*, 39-54.

Clark, E. (1984). Housing policies and new construction: A study of chains of moves in southwest Skane. *Scandinavian Housing and Planning Research, 1*, 3-14.

Hegedüs, J. (1987a). Reconsidering the roles of the state and the market in socialist housing systems. *International Journal of Urban and Regional Research, 11*, 79-97.

Hegedüs, J. (1987b). *Structural changes in the Hungarian housing market in the 70's.* Paper prepared for the International Housing Conference, Glasgow.

Hegedüs, J. (1988). Self-help housing in Hungary: Hypotheses on the changing role of the private sphere in the housing system. *Trialog, 18*, 31-36.

Hegedüs, J., & Tosics, I. (1983). Housing classes and housing policy: Some changes in the Budapest housing market. *International Journal of Urban and Regional Research, 7*, 467-494.

Hegedüs, J., & Tosics, I. (1988). *Filtration in socialist housing systems: Results of vacancy chain surveys in Hungary.* Unpublished manuscript.

Konrád, G., & Szelényi, I. (1977). Social conflicts of under-urbanization. In M. Harloe (Ed.), *Captive cities* (pp. 157-175). London: John Wiley.

Kornai, J. (1983). *Economics of shortage.* Amsterdam: North-Holland.

Lowe, S., & Tosics, I. (1988). The social use of market processes in British and Hungarian housing policies. *Housing Studies, 3*, 159-172.

Pickvance, C. G. (1988). Employers, labour markets and redistribution under state socialism: An interpretation of housing policy in Hungary 1960-1983. *Sociology, 22*, 193-214.

Sharpe, C. A. (1978). New construction and housing turnover: Vacancy chains in Toronto. *Canadian Geographer, 22*, 130-144.

Sik, E. (1988). Reciprocal exchange of labour in Hungary. In R. E. Pahl (Ed.), *On work.* Oxford: Basil Blackwell.

Szelényi, I. (1983). *Urban inequalities under state socialism.* Oxford: Oxford University Press.

Szelényi, I. (1987). Housing inequalities and occupational segregation in state socialist cities. *International Journal of Urban and Regional Research, 11*, 1-8.

Szelényi, I., & Manchin, R. (1987). Social policy and state socialism. In G. Esping-Anderson, L. Rainwater, & M. Rein (Eds.), *Stagnation and renewal in social policy.* White Plains, NY: M. E. Sharpe.

Tosics, I. (1987). Privatization in housing policy: The case of the Western countries and that of Hungary. *International Journal of Urban and Regional Research, 11*, 61-78.

Tosics, I. (1988). Inequalities in East European cities: Can redistribution ever be equalizing, and if so, why should we avoid it? A reply to Iván Szelényi. *International Journal of Urban and Regional Research, 12*, 133-136.

15

Housing Provision in Yugoslavia: Changing Roles of the State, Market, and Informal Sectors

SRNA MANDIČ

PRIVATIZATION HAS BEEN RECOGNIZED as a prominent trend in the housing policies of advanced capitalist societies. It is described in terms of a general shift of responsibility for housing provision from the state to the market. The increasing reliance on private financing, private construction, and private ownership of the housing stock starting in the 1970s corresponds to governments' gradual retreat from the type of direct involvement characteristic of the massive production programs of the 1950s and 1960s, in which government subsidies to producers successfully stimulated the production of housing, thereby reducing the severe housing shortages of the early postwar period. Since by the 1970s these shortages were considerably alleviated, governments increasingly assigned the responsibility for housing production to the private market while adjusting the instruments for their support. Instead of direct producer subsidies, governments turned to consumer subsidies, providing housing allowances to renters and tax concessions to home owners. These indirect forms of housing subsidies, the center-piece of privatization, have not only led to a marked increase in home ownership, but have also challenged traditional notions of left and right, having an appeal to parties of different ideological orientations (Adams, 1987).

AUTHOR'S NOTE: This chapter is a fully revised version of my paper "Modes of Housing Regulation and a Yugoslav Blind Spot," written for the International Research Conference on Housing, Policy, and Urban Innovation in Amsterdam, June 27-July 1, 1988. I would like to thank Mr. W. van Vliet-- and Mr. J. van Weesep for their critical remarks and suggestions.

Lately, privatization is also becoming a topic in discussions of housing policies in Eastern Europe. The trend of privatization is associated with housing reforms aimed at decreasing direct state intervention in the provision of housing (Hegedüs & Tosics, 1989; Tosics, 1987). Does the term *privatization in housing policy* apply to the same phenomena in Eastern European countries as it does in advanced capitalist societies? If common features are identified in the housing policies of both kinds of societies, do they necessarily have the same meaning and same social consequences? Could they be similarly evaluated? Tosics (1987) argues not only that in Hungary the steps taken toward privatization were introduced in wholly different circumstances, due to the different housing systems, but that they also imply different consequences, some of them being particularly beneficial. He concludes that "privatization tendencies are misleadingly similar, in fact they play an entirely different role in the west than they do in Hungary."

The trend toward privatization should thus be regarded with caution since it applies to processes that are highly context dependent. This refers not only to a possible explanation of the dynamics of privatization and to evaluation of its effects within a particular society, but also to its mere description, since its manifestation may take different forms. Differences may occur because of the fact, for instance, that in socialist economies *market* may refer not only "to an entirely different concept in several respects" (Tosics, 1987, p. 75), but also to different institutions involved as market or state agents. The problem of different meanings being assigned to terms like *state* and *market* has been recognized in some of the latest discussions in which participants from both the West and the East were present.[1]

Yugoslav housing policy of the late 1970s and thereafter has undergone a number of changes, tending toward greater promotion of home ownership. This chapter deals with the trend toward privatization in Yugoslav housing policy as manifested in the changing roles of the state, social, market, and informal sectors in housing. It focuses on a description of the trend and its principal social effects within the Yugoslav social context. Specific for this context is the prominent role of the social sector and social ownership, different from both state and private ownership. This form of ownership is lately in Yugoslavia increasingly losing its credibility, yet it has had a tremendous impact on the system of housing policy, creating a vast institutional structure hardly comparable to other systems.

PRINCIPAL FEATURES OF
YUGOSLAV HOUSING PROVISION

In this section I will describe how housing production and consumption came to be organized in Yugoslavia by the mid-1970s, the starting point for further discussion. As the state gradually retreated from the provision of housing, enterprises were assigned an increasingly important role in this sphere.

THE ROLE OF THE ENTERPRISES
IN HOUSING PROVISION

It should be pointed out that private enterprises, limited in their size and restricted to only certain kinds of goods and services, have retained an extremely marginal role, while the economy was dominated by enterprises of another type, which were transformed, in the official term, from *state enterprises* of the early 1950s into *enterprises* of the 1960s, to become later *basic organizations of associated labor.* The last two types are referred to as the *social sector* and *social ownership.*

In the earliest postwar period housing provision was mainly a state activity. With the aim of reducing the quantitative shortage, the state produced mass, collective, low-cost, low-standard housing. It was constructed by the state building enterprises, financed by the state budget, and administered as well as allocated by the state apparatus. In 1953 the state housing provision was decentralized and the central state agencies gave way to republic and municipal housing boards, which were only partly dependent on the central state budget.

In 1965 a radical economic reform was introduced to provide for more economic incentives. A centrally planned economy was to be largely replaced by market mechanisms. Enterprises, adopting self-management, were granted substantial autonomy in economic decision making and also in allocation of a part of their income. These changes have led to reorganization of housing as well, since the responsibility for housing provision was transferred from the state to enterprises, in several respects.

First of all, in the sphere of housing production, building enterprises, no longer owned and directed by the state, were granted relative autonomy in their production, thus becoming the supply side of the market of housing commodities *in nascemdi*. Second, as the state housing funds were abolished, the provision of housing finances was transferred to commercial banks, also emerging in this period, and to enterprises in general. Enterprises, having become able to dispose of a part of their income freely and devote it to different forms of consumption by employees, produced lending funds for their employees. Loans were granted for individual purchase or individual housing construction. Third, enterprises were assigned responsibility for rental accommodation of their employees. This was far reaching because the rental provision was institutionalized as an employment benefit, not as a legal commercial service, which was regarded as incompatible with self-management. The shift was regarded as democratization, since creation and allocation of housing resources became subject to self-management decision making inside enterprises.

Since rights of tenants were highly protected by legislation, rents determined by the municipal housing board, and the rental stock maintained by a municipal agency, the actual ability of an enterprise to dispose of its housing stock was very limited. These features of enterprise rental housing, popularly referred to as social housing, remain the same today.

With the new constitution in 1974 substantial new changes were introduced. The *enterprises* of the 1960s were transformed into *basic organizations of associated labor* and defined by the concept of social ownership,[2] elaborated by Kardelj (1972). The term *social ownership* refers to the integration of micro- and macro-level decision-making processes: Self-management was no longer confined to the internal affairs of an enterprise, but was assigned the role of the "grass-roots" element in the arena of politics in general[3] and in some specific areas, notably welfare. Here *self-managing interest communities* were established, as the model of organization in which the decisive role of enterprises in defining the welfare programs and their direct financing was to be promoted, while that of the state administration and the state budget was avoided. The model was also applied to housing, and the *self-managing interest communities for housing* (SHCs) were formed that still exist today. Through SHCs the role of enterprises in housing provision was to extend into the domains of welfare and the market.

First, the welfare dimension of housing policy, largely neglected in the previous period, was to be set on a self-managing basis and to become assigned to SHCs. Thus inside a municipality enterprises elect their delegates for the Assembly of the SHC, which accepts the programs of support for those unable to attain proper housing through employment benefit (retired, unemployed, low-income population, and employed in low-profit enterprises with poor housing funds). Basically, with some regional variability, two programs of support are carried out. By the program of *solidarity*, "solidarity flats" are produced and allocated to those in need. By the program of *mutuality*, a lending fund is produced to provide housing loans for individuals and enterprises' housing funds. The programs and the needed finances are determined by the SHC's Assembly and carried out by the SHC's professional staff. Once the plan is accepted and the needed finances determined, financial contributions of all enterprises become obligatory and legally reinforced.

Another basic function of the SHCs was to provide the institutional framework for *socially directed housing construction*, a term referring to the intention of superseding the "blind forces of the market." Through their delegates in SHCs, the demand and the supply side are to be coordinated through a mutually agreed-upon long-term plan of housing construction and renovation inside a municipality. Besides this, both sides are to be precoordinated also in the process of construction of a particular housing estate. Through such an organizational scheme anonymous and powerless purchasers of dwellings in the market should be substituted by an organized group of their representatives, of both individuals and enterprise housing boards. By paying a part of construction costs in advance they will be assigned the role of investors, and this should enable them to control the costs and the time of construction, thereby promoting greater efficiency in the building industry. Also, productivity is expected to rise through large-scale production.

THE DISTRIBUTION OF TENURE AND HOUSING OPPORTUNITIES

The tenure structure in the mid-1970s and also today is basically bipolar, consisting of social renting and owner occupation. Statistical data on the tenure structure are not entirely reliable, because they are based on the population census. According to Savezni Zavod za Statistiku (SZS; 1973, p. 509), in 1971, 82% of the total housing stock was privately owned. The social rental stock—that is, "solidarity dwellings" and those of enterprises—represented 18%, amounting to 35% in urban settlements. If, concerning the dominant form of tenure, societies may be classified into home owning and cost renting (Kemeny, 1981, pp. 41-64), Yugoslavia has been an example of a society with a cost-renting ideology and home-owning practice.

There is no cooperative rental accommodation, since cooperatives have been confined to the domain of building activities only. In larger cities, private profit renting has evolved in the form of subletting. It is operated on individual terms, since private ownership was legally restricted to only two housing units. Minimally regulated legally and providing subtenants with practically no protection, private renting has only a marginal role, estimated to comprise 4% of the total stock in 1981.

Low rents and extensive legal protection of renters and their families have made social renting the most advantageous type of accommodation. Access to home ownership by purchase or self-help construction is considerably more costly. Yet rental housing, the preferred type by the majority of the population, has remained much too scarce to satisfy the demand of those who are in need and, according to criteria, also qualified for it.

The criteria for allocation of solidarity dwellings have been set according to the principle of need, while the enterprise rental accommodation is allocated by two principles, according to the need of claimants and their merit. Because of the lack of systematic data and rather loose definitions of criteria it was left to Yugoslav urban sociologists to identify the applied allocation principles and the social groups that benefited the most. It was pointed out by several observers that it has been the upper social strata, those with the highest professional qualifications and in the highest position in the administration body of an enterprise, who were favored by allocation of rental dwellings. This has been the dominant type of housing of the highest strata and home ownership of the lowest strata (Seferagič, 1977; Vujovič, 1980). These allocational principles have not only substantially differentiated housing opportunities of different social strata, but also provided an additional indirect mechanism of generating further social inequalities, because they imply different costs for housing. The program of solidarity, intended to provide equal opportunities for rental housing, is too limited in scope to fulfill its function, since it has reached only a marginal proportion of 1% of the Yugoslav population (Mandič, 1989a).

These features are very similar to those under state socialism: There is a permanent shortage of rental housing, declared to be a right, not a market commodity; the principles of need and merit are applied in their allocation, favoring the higher social strata; poorer groups, to be housed, pay more in terms of money and labor; allocation of privileges and costs through the housing economy has generated not only housing inequalities between different strata, but also increased effective income inequality (Szelényi, 1983). And still in Yugoslavia these features are generated through entirely different institutional arrangements, characterized by a considerable retreat of the state. The case of Yugoslav housing provision thus seems to confirm Szelényi's argument, that these features derive from the logic of policies of equality applied in conditions of scarcity.

Yet there is also another generator of housing inequalities that may be attributed to the market. An individual's chances for housing have also been dependent on his or her branch of employment, since housing funds of enterprises in high-profit branches have been providing their employees with better housing opportunities. Also, the income of the population earned in the informal economy has an important impact on housing opportunities.

On the other hand, the state, self-management, and political organizations' apparatuses have been providing their employees with good housing opportunities. Representing 4.1% of the total employed population, this group has obtained 8.3% of the total of rental dwellings distributed in 1978 (Jugoslovenski institut za urbanizam i stanovanje [JIUS], 1987a, p. 20). This employment benefit has attracted a part of a *critical intelligentsia,* willing to demonstrate their political loyalty during a couple of years of employment in exchange for a rental apartment or a substantial housing loan.

Yet another source of inequalities in housing opportunities is the access to loans for either purchase or individual building. While the enterprises have been granting loans according to the same allocation principles as the rental accommodation, the banks have applied another selective criterion. Since mortgages have not been introduced, loans have been granted on the basis of individuals' savings, and a certain amount of down payment has been requested. Among households that have either purchased or built their homes, only half have ever taken out loans. In the group with the highest level of education, loans were used by 70%; in the group with the lowest education, 34% used loans (Mandič, 1989b). Thus a substantial number of middle- and lower-income families have been left without an opportunity to benefit from loans.

EMERGENCE OF THE CURRENT CRISIS

A housing crisis began at the end of the 1970s and was prolonged and intensified in the 1980s. A first indicator is that input into housing was drastically reduced after 1976. The proportion of housing investment in GNP decreased from 7.2% in 1976 to 6.4% in 1981, decreasing still further to 4.4% in 1985 (JIUS, 1987a, p. 40). Input reduction in housing is noticeable in enterprise housing funds, which collect fewer funds. The reduction was also due to the change in the legal system of income allocation in an enterprise, leaving it less to dispose of and at the same time lowering the priority position of housing funds. There was also a reduction in bank lending capital. In the period from 1981 to 1986 this and enterprise lending capital were reduced by half, from 4.8% to 2.4% of GNP (JIUS, 1987b, p. 12).

Public housing finances were not only reduced but also restructured as to their purpose. Enterprise housing funds and the SHCs' "mutuality funds" were increasingly used to provide housing loans for home ownership, while finances for the provision of rental housing were reduced. From 1974 to 1983, the number of loans from enterprises[4] has increased by 54% and the number of allocated rental dwellings by 9.4%, while the number of the employees increased by 38% (Kujovič, 1985). Yet the increasing number of housing loans does not imply better access to lending capital. Facing high demand and decreasing finances, enterprise housing boards simply tend to divide the available finances into smaller pieces. This trend was promoted by housing administration officials, who claimed that "the responsibility for housing provision has to shift *from the society to the individual*." This phrase has become the official slogan of housing policy, aiming to increase an individual's share in financing of housing. In the beginning of the 1980s, also, "financial participation" for obtaining a rental dwelling was introduced.

Second, the reduction of input into housing was followed by a reduction of output. The level of housing construction was highest in 1976, with 7.0 newly constructed dwellings per 1,000 inhabitants and, according to SZS (1987, p. 298, 212), decreased to 6.1 in 1980 and further to 5.5 in 1985.

Third, reduced input into housing is an important external cause for the current crisis, but there were and still are important internal factors, notably inefficiency of housing construction. It became apparent by the early 1980s that the results of socially directed housing construction were contrary to expectations. The exercise of control over the costs and time of construction required extreme involvement on the part of the "organized demand," as well as the highest-level authorities and the Communist party leaders (Kos, 1989). As this was the case only exceptionally, in practice the prices of new construction continued to rise.

Coinciding with the crisis was the trend toward privatization in housing. Two of its features were the same as in Western European societies: reduction of public financing and its restructuring toward the promotion of home ownership. Yet these two features are considered as the "second step" in privatization, occurring only after the first phase, in which quantitative housing shortages were eliminated and effective housing demand was created by private lending institutions and tax allowances (Tosics, 1987, p. 64).

In Yugoslavia social circumstances were very different. Quantitative housing shortages were far from being eliminated. In the early 1970s the Yugoslav average was 0.69 rooms per person, less than in Czechoslovakia, Hungary, and Poland, where the corresponding figures were 0.92, 0.80, and 0.74, respectively. According to SZS (1984, p. 539), in 1981 only 52% of housing units had a bathroom. Yet another substantial difference was that conditions for an effective demand were not created, and the purchasing power remained very low. Not only did lending capital decrease, but the real value of wages decreased, by 19% in the period from 1981 to 1986 (JIUS, 1987b, p. 22). Given the decrease of both wages and the efficiency of the building industry, an average-size new apartment came to be worth 11.79 average annual incomes in 1980, while in 1970 it was 6.95 (Klemenčič, 1985, p. 643).

Was there any subsidy for home ownership? Since a personal income tax has never been introduced in Yugoslavia, tax allowances were not an available instrument for subsidization of the demand. Yet the state has allowed for another indirect "instrument": inflation. It rose from 9% in 1976 to 30% in 1980, 79% in 1985, and 130% in 1987. In this period the interest rates for lending capital were kept from 4% to 10%. While the lending terms of enterprise housing loans may be attributed to self-management decisions, the lending terms of banks were the responsibility of the state. When in 1987 the bank interest finally became adjusted to inflation, 90% of their existing total lending capital had already been granted by fixed interest rates below 10% for a period of from 15 to 25 years (JIUS, 1987a, pp. 60-61).

It was thus the lending system that served a mass indirect subsidization of purchasers of new dwellings and was providing the inefficient building industry with at least some demand. But the effect was only temporary, since the lending capital was not reproduced. In the early 1980s only 30% of the real loan value had been returned to the lender (Andreevski, 1982). The popularly used term *privatization* in Yugoslavia refers to this leakage of social sector finances into private wealth. Housing loans have been one of the dominant mechanisms by which the upper strata have benefited from inflation.

Yet another difference from Western privatization was an already very low rate of rental dwellings—20% of the total stock. The scarcity of rental stock was intensified by the lack of mobility, notably from renting to home ownership. Privatization did not promote such mobility, since tenants had no incentive to leave rental dwellings. Contrary to

Hungarian efforts in promoting a filtration process and a turnover (Hegedüs & Tosics, 1988), the instruments for launching housing mobility were not considered in Yugoslavia.

MAINTAINING THE CRISIS

The trend toward privatization of housing provision was promoted through the same institutional structure that had been created earlier, only there was a considerable reduction of investment into rental housing, reinforced by the new ideology, according to which the responsibility for housing provision was shifted "from the society to individuals." Given the lack of incentives for housing mobility, however, the shift does not affect the "already housed" part of the population, but only those to be housed under new circumstances. Besides lesser chances for rental accommodation, they are also facing devastated enterprise lending funds and, since 1987, also very expensive bank loans. Thus the 1980s introduced an intergenerational bias in housing opportunities. A new pattern of housing shortages emerged in which the younger urban population, unable to enter either the rental sector or the market of housing commodities, has a notable part. Findings of the 1987 Ljubljana Housing Conditions Survey sample of the employed population, aged 18 to 35, found 67% with unsuitable housing.[5] Some 44% of the sample have declared dissatisfaction with their recent housing and no prospects to attain more suitable accommodation. One-half of the sample claimed that the lack of housing opportunities influenced their decisions on family size (Mandič, 1989b). Yet in Ljubljana and in the Republic of Slovenia the general housing standard is among the highest in Yugoslavia.

Since the system of housing policy has not sufficiently provided for the institutional supportive environment that would correspond to the market type of housing provision, housing provision is increasingly relying on other supportive institutions in the informal sector.

THE INFORMAL SECTOR

The informal sector was traditionally well developed, especially in rural areas, and has been in operation throughout the entire postwar period, yet it is still further encouraged by the current housing crisis. Informal housing provision involves activities and transactions of housing resources outside the formal institutions of housing policy. A brief description of the informal sector's institutions and activities follows.

The basic institution of informal housing provision is a kinship network, with extensions to other informal relations. This kind of organization proves to be the most effective and rational form under circumstances of housing resource shortage because it provides an effective mechanism of multiplication of an individual's resources. These networks also provide an intergenerational flow of resources by which the younger generation can partly compensate for disadvantages created by the formal system. There are a number of activities in which the kinship network can help an individual: by cash or loan in case of a purchase, by information and informal help in access to rental stock, by inheritance of the property or the rental tenure, and, above all, by shared use of dwellings. In the above-mentioned sample of the employed population of Ljubljana, aged 18 to 35, one-third was still living with parents (Mandič, 1989b). This type of housing provision is a hidden form of homelessness, and has an important effect on the process of housing needs articulation: Conflicts arising from overcrowding and lack of privacy tend to be articulated in terms of interpersonal relations rather than housing policy criticism. A second very important role of kinship is in self-help construction. It provides for an even broader scope of kinship resources: land, building materials, finances, skills, information, and, above all, labor. The self-help construction enables relative shortages of resources to be compensated for with labor.

To estimate the extent of both formal and informal systems of housing provision, in the 1987 Quality of Life in Yugoslavia survey a distinction between public and private (kinship) resources was used. In the sample of the adult population, only 42% have ever used public resources, bank loans included. Half were dependent exclusively on kinship resources (Mandič, 1989a). The formal model of housing provision, incorporated in the housing policy, thus appears to be limited in its performance and leaves it to the informal model to compensate for its ever larger dysfunctions.

Yet the informal housing provision, based on kinship resources, is only one, nonmonetarized, part of the informal sector. The other, monetarized, part has also been very important. Because private entrepreneurship has been restricted to only some fragments of the construction process and is also subject to excessive taxation, a substantial part of its activities has remained officially unregistered, illegal. It provides self-help construction with services and also provides a considerable part of the population with opportunities for additional, unregistered, earnings. The phenomenon exists also in other socialist countries; in

Hungary it is referred to as the "second economy" (Hegedüs & Tosics, 1988).

CONCLUSIONS

I have tried to identify the principal features of the changes that occurred in housing provision in Yugoslavia during the late 1970s and the 1980s. Some of them, notably the reduction of public finances and their restructuring toward promotion of home ownership, are very similar to those of Western-type privatization. Yet I have pointed out that in Yugoslavia these changes were introduced in different circumstances. The most important one was that the conditions for an effective demand in the housing market were not created either in terms of high earnings or in terms of the available and affordable lending capital. Is this difference an important element in the explanation of the trend toward privatization?

According to Harloe (1981), it is precisely the high, stable earnings of a substantial proportion of the population that are a vehicle of the current increase in the rate of owner occupation in advanced capitalist societies. Harloe explains the phenomenon in terms of the process of recommodification, referring to the long-term transition of commodified housing provision from one form to another; from private rental, dominating in the early part of the century, to the currently dominant, "modern and (from the point of view of capital) most effective form— owner occupation," while the intermediary period in which social housing dominated is regarded as a period of transition.

From this perspective the trend toward privatization in Yugoslav housing policy cannot possibly be regarded as recommodification. Home ownership here is not the latest, modern form of commodification, but, under a socialist economy, its prime, original form, partly resulting from the scarcity of rental accommodation, which is excluded from commodification.

It has been suggested by Pickvance (1986, p. 165) that the explanation of the level of owner occupation requires two diverse explanatory models. For advanced capitalist societies the explanatory variable is "the level of stable earnings," as defined by Harloe. For the state socialist societies it is the decrease of state house building.

The case of privatization in Yugoslav housing policy clearly demonstrates that reliance on the Western explanatory model would be misleading and thus shows that Pickvance is right in introducing diverse explanatory models. Yet the decrease of state house building does not seem to be an explanatory variable for the Yugoslav case of privatization. Earlier in this chapter I pointed out how housing provision in Yugoslavia is in many respects similar to that in state-socialist countries and how this cannot be attributed to the same role of the state. In Yugoslavia the responsibility for housing provision has shifted from the state to another institutional arrangement, to self-managing interest communities for housing. If there is a considerable variability in the system of housing policy in Hungary and Yugoslavia, it seems that there is much less variability in the mechanisms that have developed as complementary to the formal system. If privatization implies the increase of the self-help provision and only very limited marketization (Hegedüs & Tosics, 1988), than Hungary and Yugoslavia have followed a very similar pattern of changes in housing provision.

NOTES

1. See the international research conferences on Housing Between State and Market held in Dubrovnik, 1988, and on Housing Reforms in Eastern Europe, held in Noszvaj, 1989.

2. Social ownership, as is today generally believed, was defined ideologically and not in classical terms of property rights. According to the concept, the social ownership of enterprise was regarded neither as the state's nor as the collective property of the employees, but as the common property of the working class as a whole. Thus the influence of the working class as a whole on the management of a particular enterprise was provided for by a detailed institutionalization (representation of the "broader social interests") and an extensively elaborated legislation, highly restricting the scope of autonomy.

3. The federal, republic, and municipal assemblies consist of representatives, elected by associated labor, lower-level territorial units, and political organizations.

4. Data refer to a large sample of enterprises included in a national statistical survey.

5. Unsuitable housing was defined as being without bathroom and/or toilet and/or a kitchenette and/or being damp and/or being overcrowded (i.e., having less than 16 m^2 of floor space per person).

REFERENCES

Adams, C. T. (1987). The politics of privatization. In B. Turner, J. Kemeny, & L. L. Lundquist (Eds.), *Between state and market: Housing in the post-industrial era* (pp. 127-155). Stockholm: Almquist & Wiksell International.

Andreevski, U. (1982). Ekonomska stanarina neophodnost—konverzija kredita imperativ. *Socijalizam, 1.*

Harloe, M. (1981). The recommodification of housing. In M. Harloe & E. Lebas (Eds.), *City, class and capital* (pp. 17-50). London: Edward Arnold.

Hegedüs, J., & Tosics, I. (1988). *Is there a Hungarian model of housing system?* Paper presented at the Conference on Housing Between State and Market, Dubrovnik.

Hegedüs, J., & Tosics, I. (1989). *Divergences and convergences in the development of East European housing systems.* Paper presented at the Conference on Housing Reforms in Eastern Europe, Noszvaj.

Jugoslovenski institut za urbanizam i stanovanje (JIUS). (1987a). *Stambena politika i stanovanje u SFRJ*. Beograd: Author.

Jugoslovenski institut za urbanizam i stanovanje (JIUS). (1987b). *Finansiranje stambene izgradnje*. Beograd: Author.

Kardelj, E. (1972). *Protislovja drušbene lastnine v sodobni socialistični praksi*. Ljubljana: Dršavna zalošba Slovenije.

Kemeny, J. (1981). *The myth of home-ownership*. London: Routledge & Kegan Paul.

Klemenčič, T. (1985). *Stanovanjsko gospodarstvo*. Ljubljana: CGP DELO.

Kos, D. (1989). *Informal activities in the formal housing system: Lessons for urban planning*. Paper presented at the Conference on Housing Reforms in Eastern Europe, Noszvaj.

Kujovič, N. (1985). Raspodela stanova i kredita za stanove. *Jugoslovenski pregled, 9*, 321-326.

Mandič, S. (1989). Prispevek k opisu stanovanjske preskrbe v Jugoslaviji. In K. Boh (Ed.), *Stratifikacija in kvalitete življenja* (pp. 1-29). Ljubljana: ISU.

Mandič, S. (1989). *Stanovanjsko vprašanje mladih v Ljubljani*. Ljubljana: Samoupravna stanovanjska skupnost Ljubljanskih občin.

Pickvance, C. G. (1986). Comparative urban analysis and assumptions about causality. *International Journal of Urban and Regional Research, 2*, 162-184.

Savezni Zavod za Statistiku (SZS). (1973). *Statistički godišnjak Jugoslavije*.

Savezni Zavod za Statistiku (SZS). (1984). *Statistički godišnjak Jugoslavije*.

Savezni Zavod za Statistiku (SZS). (1987). *Statistički godišnjak Jugoslavije*.

Seferagič, D. (1977). *Socijalna segregacija u rezidencijalnom prostoru*. Zagreb: Filozofski fakultet Sveučilišta u Zagrebu.

Szelényi, I. (1983). *Urban inequalities under state socialism*. New York: Oxford University Press.

Tosics, I. (1987). Privatization in housing policy: The case of the Western countries and that of Hungary. *International Journal of Urban and Regional Research, 1*, 61-78.

Vujovič, S. (1980). Pojmovno-hipotetički okvir za istraživanje stanovanja. In *Porodica i društveni sistem*. Beograd: Institut za socijalnu politiku.

Conclusion

Directions in Housing Research

JAN van WEESEP

FOR MANY YEARS, housing research was concerned with the development of architectural styles. Most studies were limited in scope, concentrating on historic landmarks or buildings characteristic of a particular period or place. A few went beyond the "biographical" approach and took a broader historical perspective, studying the evolution of residential construction, relating form to function. Since then, the field has come a long way from this fascination with the morphology of the built environment.

In his seminal work, Smith (1971) argues that a dwelling can provide four distinct functions: shelter, privacy, location, and investment. The importance of each of these varies among cultures and from one period to the next. There is an implicit ranking in this series: Once a household has met the basic need for shelter, the privacy component is considered, and so on. The functions also define fields of study, which continue to develop: The frontier of housing research has moved beyond the shelter issue (who lives where, under what circumstances?) to the analysis of the social and economic functions of housing. The recent flurry of research on homelessness—intrinsically a shelter issue—is indicative of this evolution; most of the work highlights the social dimension of the problem.

As the topical issues changed, researchers shifted from description to explanation, focusing on differences in housing situations and the dynamics of the housing system. These explanations initially concerned the distribution mechanisms of housing markets. Subsequently, external factors were brought into the analyses. Differences in housing are now commonly accounted for by referring to their demographic, economic, social, technological, and spatial contexts. But above all, the attention of social scientists has been captured by the links between housing and developments in society at large.

The political arena is increasingly subjected to scrutiny in search of explanations for change in housing situations. The frequent initiatives for new policies in themselves provide challenges for evaluation research. Thatcherism, Reaganomics, and the retreat of government in many other countries—of which the dismantling of the monolithic regimes of Eastern Europe provide a dramatic recent example—have had far-reaching effects on the provision of housing and on the way it performs its social functions. The trend throughout the social sciences to study decision making in light of changes in the social structure has amplified the wave of contextual and policy-oriented studies. By focusing on aspects of the processes of privatization and decentralization in the field of housing, the present volume reflects this concern of recent research.

In this brief epilogue, some of the most viable strands of recent research are reviewed. First, some comments are made on the still expanding field of cross-national comparisons. This is followed by an argument in favor of assessing housing policy in the context of wider policy fields. This paves the way for a discussion of the need to analyze housing as a structuring force in society. The chapter concludes by advocating an improved organization of this field of research.

ON COMPARATIVE HOUSING RESEARCH

After a period in which housing studies were limited to the description of unique elements of the "native" setting, cross-national studies flourished and produced an extensive body of literature. Bourne's (1986) assessment of the comparative urban literature seems to apply to some of these studies as well: tour-group research and hastily compiled anthologies, along with elementary analysis, superficial interpretations, and misleading generalizations. In their discussion of comparative urban studies, Walton and Masotti (1976) caution against misinterpretation resulting from ventures into "foreign" settings with "prefabricated methodological and theoretical tools, which presuppose the order and meaning of events."

Practical difficulties may hinder the collection of data, but conceptual difficulties can invalidate the results of the work. Not only do definitions differ among countries (van Vliet--, 1989), but specific functions of housing may be met in entirely different ways in other times and places. Not so long ago, the private rental sector in Britain provided housing of last resort (Kemp, 1988), but lately the social rental sector shows signs of becoming marginalized (Forrest & Murie, 1986). In the Netherlands, parts of the owner-occupier sector provided refuge for ethnic minorities and recent immigrants barred from social rental housing (van Hoorn & van Ginkel, 1986), while young people created a new (short-lived) sector by squatting vacant properties (van Weesep, 1984).

Yet the growing number of thoughtful descriptions of the state of housing in various countries demonstrates how national studies can contribute empirical content to comparative analysis; such studies add depth of interpretation and assist in the search for explanations. Such books—like the present one—can demonstrate the range of issues that form part of a particular cross-national trend. Local experts can be involved, so that the pitfall of misinterpreting phenomena can be avoided. In addition, such collections of studies may provide inspiration for policy.

In the next few years, the focus of comparative studies must shift to unleash their potential. Pickvance (1986) has argued that there is a crucial distinction between the more descriptive comparative research and comparative analysis, which attempts to understand two or more cases in terms of one or more models. Such explanation can incorporate a wider range of variables than is possible in a study of one single country, such as characteristics of national housing systems—including national policy instruments. The influence of national variations can thus be evaluated and the researcher can distinguish among broad patterns of social change, government policy, and institutional restructuring (Ball, 1988). New policy ideas can be generated and the strong and weak points of national housing systems and policies can be evaluated. In studies of a single country the impact of policy and restructuring can be approximated only by historical analysis under the *ceteris paribus* condition or in simulation studies. Researchers should debate the theoretical underpinnings, refine the methodology, and create the organizational conditions to advance to this stage in comparative study.

HOUSING STUDIES
IN A BROADER CONTEXT

The idea that the provision of housing is a social issue is not new. By the end of the nineteenth century, social researchers and policymakers shifted their focus from housing situations to the nature of poverty. The inability of specific population categories to obtain decent housing became their major concern. This ideologically inspired shift eventually laid the foundation for a new set of policies. The tremendous expansion of the social rental sector after World War II was a clear expression of the general trend toward increased government involvement in social welfare. Similarly, the retreat of government in recent years reflects the preference of political conservatives to give direct support only to people whose need can be established unequivocally (Lundqvist, 1986). As the studies in this book reveal, this change is expressed in many policies. Yet everywhere it entails less support for "brick-and-mortar" subsidies and more attention to income support programs. Clearly, housing policy is increasingly considered to be merely one of the elements of welfare policy. This makes it vulnerable to the forces that have recently revamped the benevolent state into a no-nonsense business operation. Researchers should pay more attention to the position of housing among the other areas of social policy. If they cannot demonstrate its special position, they will not be able to help defend housing from the onslaught of budget cuts.

Whereas the privacy function of housing is related to the match of household characteristics and housing needs to the features of the dwelling, the locational function concerns its external utility. The location of each residence is fixed and determines access to work, services, and facilities. Location also provides each dwelling with a residential environment that has positive or negative externalities. This is not an aspect of housing research only; the nature of the relationship between the built environment and its users, as well as between patterns of land use and social structures, puts housing studies at the core of urban research. Social conflicts, such as those inherent in neighborhood change, can be explained in terms of the competition for scarce locations. The finite supply of suitable and desirable locations is directly related to conflict, for instance, in the zero-sum trade-off between urban revitalization and urban renewal policies. By connecting the analysis of housing systems to urban dynamics, researchers can provide a stimulus to the understanding of urban change and the design of appropriate and effective policies. At present, the debate in housing is focused on the analysis of housing policies in the context of general shifts toward privatization and decentralization of authority from central to local governments. In the coming years, housing research should take a wider perspective, and analyze housing issues in an even broader social context. The relationships between housing market and labor market, between residential location and use of urban services and facilities, between residential construction policy and urban policy, for example, offer many challenging and rewarding research topics.

HOUSING AS A STRUCTURING FORCE

In recent publications, Saunders (1984, 1989) has explained how people's roles and activities as consumers are crucial to their social positions and identity. He considers the home as the main object of consumption, as well as a container within which much consumption takes place. Thus housing is a major structuring force in society.

A rapid inflation of real estate values can have severely debilitating effects on a community. Individual home owners can profit from this rise in value, but the increase in their capital assets can create social cleavages in otherwise homogeneous groups. It is not yet clear how the value increases are used, nor how the resulting changes in consumption potential vary among households and among residential areas and regions of the country. Such differences become intergenerational if the increase in equity is passed on to the next generation, either when children embark upon their own housing careers or by inheritance. Unfortunately, most of the work in this important field of study is still speculative, and a major effort should be made to collect and analyze hard data.

The effect of housing subsidies on income distribution has attracted widespread attention. But the diversity of programs and the restrictions on access to the data (to safeguard privacy) make this topic difficult to investigate. Yet, some have succeeded (e.g., Flood & Yates, 1989). They show that while the ideological changes have targeted the programs more effectively to those in need of help, the indirect subsidies still reflect older priorities. When all forms of subsidy are taken together, higher-income groups benefit most because of tax advantages and other financial policies.

Systematic research is lacking in this field. Thus the frequently voiced arguments to promote specific forms of tenure stand on shaky ground. More research on housing subsidy programs is urgently needed, in view of the suggested links between (housing) consumption positions and the shifting structure of society. The empirical work must draw on many sources; in a recently published example, Byrne (1989) used secondary analysis of census data and housing management records, survey findings, historical analysis of documents, and interviews with local informants. The analysis of such diverse material is demanding, but Byrne demonstrates that it can shed light on otherwise hidden facets of sociotenurial polarization.

THE NEED FOR ORGANIZATION IN THE RESEARCH FIELD

In recent years, several major conferences have been devoted to assessing advances in housing research and the exploration of new topics, theories, and methodologies. The research conference on Housing, Policy and Urban Innovation, which convened in Amsterdam in 1988, followed two earlier stock-taking conferences in Amsterdam (1985) and Gavle, Sweden (1986). Follow-up conferences are planned for 1990 and 1992.

Such meetings provide a forum for the presentation and discussion of ideas and research findings. Indirectly, they stimulate the development of housing studies by disseminating results among the participants and by the publication of proceedings. Yet, to develop their full potential, housing studies need the concerted effort of researchers *in the field*. Rather than organizing such broad conferences even more frequently, housing specialists should organize networks and working groups. Such working modules can discuss research priorities and direct their members' research efforts. Until now, the development of the field has depended on the resourcefulness of individual researchers. While this may not have had an adverse effect on the formulation of ideas and hypotheses, it has clearly hindered theory development and the construction of a broad, pertinent empirical base. Given the difficulties of cross-cultural work elaborated above, the field of housing studies would clearly benefit from a concerted research effort. Also, the development of policy would reap the benefits of improved theoretical insights: Good theories do not necessarily generate good policy, but at least they can help defeat bad policy.

REFERENCES

Ball, M. (1988). Housing provision and comparative housing research. In M. Ball, M. Harloe, & M. Maartens (Eds.), *Housing and social change in Europe and the USA*. London: Routledge.

Bourne, L. S. (1986). Urban policy research in comparative perspective: Some pitfalls and potentials. *Tijdschrift voor Economische en Sociale Geografie, 77*, 163-168.

Byrne, D. (1989). Sociotenurial polarization: Issues of production and consumption in a locality. *International Journal of Urban and International Research, 13*, 369-389.

Forrest, R., & Murie, A. (1986). Marginalization and subsidized individualism. *International Journal of Urban and Regional Research, 10*, 46-65.

Flood, J., & Yates, J. (1989). Housing subsidies and income distribution. *Housing Studies, 4,* 193-210.

Kemp, P. (1988). *The future of private renting.* Salford: Environmental Health and Housing Division.

Lundqvist, L. J. (1986). *Housing policy and equality: A comparative study of tenure conversions and their effects.* London: Croom Helm.

Pickvance, C. G. (1986). Comparative urban analysis and assumptions about causality. *International Journal of Urban and Regional Research, 10,* 162-184.

Saunders, P. (1984). Beyond housing classes. *International Journal of Urban and International Research, 8,* 202-227.

Saunders, P. (1989). The meaning of "home" in contemporary English culture. *Housing Studies, 4,* 177-192.

Smith, W. F. (1971). *Housing: The social and economic elements.* Berkeley: University of California Press.

van Hoorn, F. J. J. H., & van Ginkel, J. A. (1986) Racial leapfrogging in a controlled housing market. *Tijdschrift voor Economische en Sociale Geografie, 77,* 187-196.

van Vliet--, W. (1989). Cross-national housing research: Analytical and substantive issues. In W. van Vliet-- (Ed.), *The international handbook of housing policies and practices.* Westport, CT: Greenwood/Praeger.

van Weesep, J. (1984). Intervention in the Netherlands: Urban housing policy and market response. *Urban Affairs Quarterly, 19,* 329-353.

Walton, J., & Masotti, L. H. (1976). *The city in comparative perspective: Cross-national research and new directions in theory.* New York: John Wiley.

Index

About the Contributors

CAROLYN TEICH ADAMS is a Professor at Temple University, where she is Chair of the Department of Geography and Urban Studies. She teaches courses in urban public policy, housing, and program evaluation, and has authored numerous articles and books, including an award-winning cross-national study, *Comparative Public Policy: The Politics of Social Choice in Europe and America,* now in its third edition. In 1988 she published *The Politics of Capital Investment,* a study of the impact of city investments on housing values in the neighborhoods of Philadelphia. She consults widely with nonprofit organizations on housing and community development, and is currently studying commercial ventures operated by nonprofit organizations.

JOHN S. ADAMS is Professor of Geography, Planning, and Public Affairs at the University of Minnesota, Minneapolis. He is a Past President of the Association of American Geographers and author of numerous publications, including the recent *Housing America in the 1980s* (Russell Sage Foundation, 1987).

RICHARD P. APPELBAUM is Professor at the University of California, Santa Barbara, where he is Chair of the Sociology Department. He has written extensively on rental housing and homelessness in the United States, and is coauthor of *Rethinking Rental Housing* (Temple University Press, 1988). He consults with state and local government organizations, as well as with homeless and low-income housing advocacy groups.

RACHEL G. BRATT is an Associate Professor in the Department of Urban and Environmental Policy at Tufts University. She received a Ph.D. from the Department of Urban Studies and Planning at MIT. She is the author of *Rebuilding a Low-Income Housing Policy* and a coeditor of *Critical Perspectives on Housing*. She is a current member of the Massachusetts Housing Finance Agency's Multifamily Advisory Committee, and from 1984 to 1986 she served on the Consumer Advisory Council, appointed by the Board of Governors of the Federal Reserve System.

ROBERT A. CARTER is Director-General of the Housing Corporation of New Zealand, the New Zealand government agency for housing policy. Previously, he was Senior Lecturer in Economics at the University of Melbourne, Australia, and he has published extensively on housing and public policy. Prior to taking up his current position, he was Deputy Director of the Ministry of Housing and Construction and was the architect of the Home Opportunity Loans and Victorian Housing Bonds schemes.

DAVID CHAMBERS is Principal Lecturer in the Environment at Thames Polytechnic, London. He is a Director of the Centre for Urban and Environmental Studies at Thames Polytechnic, and is currently involved in multidisciplinary research into local state policies and central-local relations.

ANTHONY D. H. CROOK is Senior Lecturer in Town and Regional Planning at the University of Sheffield, United Kingdom. His major research interest is in private rented housing. He is also involved in housing practice as a board member of nonprofit housing associations and has served as a council member of the National Federation of Housing Associations. He is currently participating in U.K. studies of the impact of the national housing finance system on six metropolitan areas and of the impact of tax concessions on the supply of private rented housing.

PETER DREIER is Director of Housing at the Boston Redevelopment Authority and Housing Policy Adviser to Mayor Ray Flynn. He previously taught sociology at Tufts University. He received his B.A. in journalism from Syracuse University and his Ph.D. in sociology from the University of Chicago. He is on the boards of the National Low-Income Housing Coalition and the National Housing Institute. He drafted the Community Housing Partnership Act and is coauthor of *Who Rules Boston? A Citizen's Guide*. His articles have appeared in the *Harvard Business Review, Social Policy, Urban Affairs Quarterly, Social Problems,* and *International Journal of Urban and Regional Re-*

search, as well as the *New York Times, Washington Post*, and elsewhere in the popular press.

MARSHALL M. A. FELDMAN is Assistant Professor of Community Planning in the Graduate Curriculum in Community Planning and Area Development at the University of Rhode Island. His research interest is in urban political economy, particularly in the areas of housing, urban spatial structure, and industrial location. His current work focuses on applying a regulation approach to industrial reorganization and regional change.

RICHARD L. FLORIDA is Assistant Professor of Public Policy and Management in the School of Urban and Public Affairs at Carnegie-Mellon University. His current research is on the role of housing in advanced industrial societies and the role of technological change in economic development.

FRED GRAY is a Lecturer in Urban Studies at the University of Sussex, Brighton, England. He is convener of the university's Continuing Education Programme. His recent publications include the coauthored *Transformation of Britain* (Fontana, 1989).

CHESTER HARTMAN is a Fellow at the Institute for Policy Studies in Washington, D.C. He has taught at Harvard, Yale, Cornell, the University of North Carolina, the University of California, Berkeley, and Columbia. He chairs Planners Network, a national organization of progressive urban and rural planners. His most recent housing books are *Housing Issues of the 1990s* (coedited with Sara Rosenberry; Praeger, 1989) and *Critical Perspectives on Housing* (coedited with Rachel Bratt and Ann Meyerson; Temple University Press, 1986).

JÓZSEF HEGEDÜS is a sociologist in the Institute of Sociology at the Hungarian Academy of Sciences, and a part-time Lecturer in Urban Sociology at Technical University of Budapest. His interests are in urban sociology, sociology of housing, and regional development. His research has included a series of vacancy studies (with Iván Tosics), a case-study series on second homes (with Robert Manchin), and an analysis of housing reforms in Hungary, 1986-1988. His recent publications include "Reconsidering the Roles of the State and the Market in Socialist Housing Systems" (*International Journal of Urban and*

Regional Research, March 1987) and "Self-Help Housing in Hungary" (*Trialog,* 1988).

W. DENNIS KEATING is Professor of Law and Urban Affairs at Cleveland State University. He formerly directed CSU's Center for Neighborhood Development. He is coauthor of a casebook titled *Housing and Community Development Law* and has participated in national assessments of community development corporations and community-based housing development. He has published widely on housing and community development policy.

MARK KLEINMAN is Lecturer in Social Administration at the London School of Economics, and a Fellow of Wolfson College, Cambridge. For the last six years he has undertaken research in the Department of Land Economy at the University of Cambridge into a number of housing and urban issues. His current research interests include investment in social housing, urban change and economic development, and comparative housing policy. He is coauthor of *Private Rented Housing in the 1980s and 1990s* (Granta Editions, 1986) and a contributor to *Beyond Thatcherism* (edited by P. Brown and R. Sparks; Open University Press, 1989) and *Social Policy Review 1988/9* (edited by M. Brenton and C. Ungerson; Longman, 1989).

NORMAN KRUMHOLZ is a Professor in the College of Urban Affairs at Cleveland State University, a position he took after a 20-year career as a planning practitioner, including serving as Planning Director for the City of Cleveland for a decade. He created and directed the Center for Neighborhood Development at CSU and teaches neighborhood planning. He served on President Carter's National Commission on Neighborhoods (1979-80), and has been President of the American Planning Association and a winner of the Rome Prize of the American Academy of Rome.

SRNA MANDIČ is a sociologist at the Institute of Sociology, University of Ljubljana, in Ljubljana, Yugoslavia, where she has been a principal researcher on a number of projects on residential segregation in Ljubljana, social housing, and housing policy in Yugoslavia. She is currently involved in comparative regional research on housing provision in Yugoslavia.

LARS NORD is an Assistant Professor of Political Science and a Senior Research Associate at the National Swedish Institute for Building Research, Gävle, Sweden, as well as Lecturer in the Department of Government, University of Uppsala. His earlier research concerned Yugoslav internal and foreign policies, socialism, and nonalignment. His current research focuses on Sweden's housing policy and includes a project on the relationship between central and local government.

CHRIS PARIS is currently Associate Professor and Director of the Australian Centre for Local Government Studies, School of Management, Canberra College of Advanced Education, after a stint as a Senior Research Fellow in the Urban Research Unit of the Australian National University. Previously, he was a Senior Researcher at the Centre for Environmental Studies, London. His recent publications include *Towards Fair Shares in Australian Housing* (with Kendig) and *Stability and Change in Australian Housing* (with Beed, Stimson, and Hugo). He is currently working on economic restructuring and urban and regional change in Australia.

KEITH P. RASEY is a mid-career Ph.D. candidate in Cleveland State University's Urban Studies Program. He has served in several federal agencies, including the U.S. Department of Housing and Urban Development, where he was Director of the Program Evaluation Division in Housing, and several presidential and departmental task forces. He was also Director of Policy and Program Development of the National Council of State Housing Agencies, and participated in a study of the impact of federal cutbacks in the Community Development Block Grant Program on local government in Ohio.

ELIZABETH A. ROISTACHER is Professor of Economics, Queens College of the City University of New York. She served as Deputy Assistant Secretary for Economic Affairs of the U.S. Department of Housing and Urban Development in 1980-81, and has written widely on housing economics and housing policy. Her publications include "Housing and the Welfare State in the United States and Western Europe" (*Netherlands Journal of Housing and Environmental Research*, 1987) and "A Tale of Two Conservatives: Housing Policy Under Reagan and Thatcher" (*Journal of the American Planning Association*, 1984).

HILARY SILVER is Assistant Professor of Sociology and Urban Studies at Brown University. Her major interests and publications are in the areas of low-income housing, local economic development, home work, the service industries, and urban inequality. Her chapter in this volume is drawn from a forthcoming book comparing American, British, and Dutch housing policies in the 1980s. She previously taught at Columbia University and has held visiting positions at New York University, the University of Sussex, and the University of Lille.

MICHAEL A. STEGMAN is Chair of the Department of City and Regional Planning at the University of North Carolina at Chapel Hill. He has written extensively on the subjects of national housing policies, housing finance, and affordable housing. His current research activities include an evaluation of the U.S. federal government's public housing privatization program, and a five-year study of the social and economic impacts of home ownership on lower-income families. From 1979 to 1981, he was a member of President Carter's administration, serving as Deputy Assistant Secretary for Research in the U.S. Department of Housing and Urban Development.

IVÁN SZELÉNYI is Professor of Sociology at the University of California, Los Angeles. Previously, he was Distinguished Professor at the Graduate School of the City University of New York (1986-88), Karl Polanyi Professor at the University of Wisconsin—Madison (1981-86), Foundation Professor of Sociology at Flinders University of South Australia (1976-81), and Research Fellow at the Institute of Sociology of the Hungarian Academy of Sciences (1963-75). He is author of *The Intellectuals on the Road to Class Power* (Harcourt Brace, 1979), *Urban Inequalities Under State Socialism* (Oxford University Press, 1983), and *Socialist Entrepreneurs: Embourgeoisement in Rural Hungary* (Polity Press, 1988), and is cowinner of the 1989 C. Wright Mills Award.

IVÁN TOSICS is a Sociologist at the Institute for Building Economy and Organization, in Budapest. His interests concern housing policy and urban sociology. His past research has examined the history of housing and the changing urban social structure of Budapest. More recently, he has focused on Hungarian housing policy reform at local and central levels. His recent publications include "Dilemmas of Reducing Direct State Control: Recent Tendencies in Hungarian Housing Policy" (in *Between State and Market: Housing in the Post-Industrial Era,* edited by B. Turner, J. Kemeny, and L. J. Lundqvist; Almqvist & Wiksell International, 1987), "Privatization in Housing Policy: The Case of the

Western Countries and That of Hungary" (*International Journal of Urban and Regional Research,* March 1987), and, with S. Lowe, "The Social Use of Market Processes in British and Hungarian Housing Policies" (*Housing Studies,* July 1988).

WILLEM van VLIET-- is an Urban and Environmental Sociologist at the University of Colorado, Boulder. His research interests concern cross-national analysis, urban and community planning, and housing. His published works include contributions to anthologies and journals in the field and a number of edited books, including most recently *Housing Markets and Policies Under Fiscal Austerity* (1987), *Women, Housing, and Community* (1989), and *The International Handbook of Housing Policies and Practices* (1990).

JAN van WEESEP is Associate Professor in the Department of Applied Geography and Planning at the University of Utrecht, the Netherlands. He is the Director of the Urban Research Program of the Institute of Geographical Research. His interest in housing focuses on the relationship of housing to urban development and social developments. His recent publications include studies of condominium conversion in the United States and the Netherlands, the sale of public housing, and housing market effects of social segmentation in cities.

ELIA WERCZBERGER is Professor of Urban Planning at the Public Policy Program at Tel Aviv University and Head of its Program in Real Estate Appraisal. His research has been concerned with the housing market, housing maintenance, mathematical planning models, multiple-criteria decision making, and environmental quality. He has published in various journals, including *Environment and Planning, Regional Science and Urban Economics, Housing Studies,* and the *Journal of Real Estate Finance and Economics.*

NOTES

NOTES

NOTES